THE WORLD
ACCORDING TO
HOMO SAPIENS

In Memory of Paul Yakovlev

THE WORLD ACCORDING TO *HOMO SAPIENS*

✦

(Or Why We Humans Experience The World The Way We Do)

Philip R. Sullivan, M.D.

iUniverse, Inc.

New York Lincoln Shanghai

THE WORLD ACCORDING TO *HOMO SAPIENS*
(Or Why We Humans Experience The World The Way We Do)

iUniverse books may be ordered through booksellers or by contacting:

iUniverse
2021 Pine Lake Road, Suite 100
Lincoln, NE 68512
www.iuniverse.com
1-800-Authors (1-800-288-4677)

ISBN: 0-595-34602-2

Printed in the United States of America

Contents

Part III   *THE ENIGMA OF HUMAN
CONSCIOUSNESS*

INTRODUCTION

As a boy, I attended Boston Latin School; and sometimes after school, I'd walk down Avenue Louis Pasteur and along the Fenway to the Museum of Fine Arts (a freebie in those days before breakneck competition had pushed art prices out of sight). One of the works that most captivated me consisted of a large canvas by Paul Gaugin depicting a group of Polynesians, their haunted facial expressions more than reinforced by their desolate postures. Perhaps, I thought, the reason for their consternation lay in the three queries that Gaugin had taken the trouble of stashing in the picture's corner: *Who are we? Where do we come from? Where are we going?*

The painting invariably evoked an eerie response from me despite the fact that I had the good fortune of already knowing the answer to all three questions. Thanks, that is, to the highly traditional Religion in which I was raised, I'd already been appraised of who I was: a child of God and heir to heaven. So not only did I come *from* God, I would eventually return *to* Him, living in beatific happiness forever. If I played my cards right. Congruently, the very word 'religion', as I'd discovered, came from two Latin roots, meaning 'to bind back'.

Struck me as strange though, even at the time, that Gaugin's painting was able to elicit such a vividly personal response. After all, given the fact that I already had a handle on the answers to life's BIG questions, I didn't have to waste time standing around like Gaugin's natives, puzzling spookily over what life was all about. But of course that's the sort of thing great artists are able to do to you. Using imaginative representations, they can evoke recognition at a gut level of complexities that simply won't give way to simple formulas.

At any rate, from that time until the present, I've tried to make sense of our *human world*. And growing up in an intellectual milieu pervaded by modern science, it's not surprising that my focus gradually shifted from contemplation of the *Supernatural* to a lifelong study of the *Natural*. After four years at Holy Cross College studying with the Jesuits, I returned to Avenue Louis Pasteur, though now to Harvard Medical School. My focus by this time had started to shift from "man, a little less than the angels" to man's place in nature, within a world of "all creatures great and small." And within that context, it was obvious that each species experiences the world in its own special way.

Let me underline a significant implication of that fact by referring to my book's title—one that I borrowed from John Irving. When he wrote his old novel, *The World According To Garp,* he shared with us the unique manner in which one particular human being viewed the world. And I think we're all agreed that we learn more about ourselves *as individuals* by comparing our own experience with that of other humans. In similar fashion, we can also learn more about ourselves *as a species* by comparing our human way of experiencing the world with that of other species. Hence, my title, *The World According To Homo Sapiens.*

If we were to indulge in fancy terminology—not to mention a bit of alliteration—we might refer to the project as falling within the field of 'Comparative Cognition'. But all we'll actually be doing is to use some examples of how simpler organisms deal with the nitty-gritty of life in order to foreshadow how our own more complex systems allow us to make sense of the world.

In the case of all animals, it's obvious that they can't even begin to deal with 'whatever' until they manage to *detect* it with their limited sensory equipment. Then they have to process the information they've managed to turn up, which immediately adds a second limitation. For animals can process information only with the nervous-system 'programs' they have available to manage the raw data. The limitations that we'll be able to see so readily in the cognitive systems of simpler animals will help us recognize more easily that our own cognitive systems are also seriously limited.

Fair trade laws nowadays often require disclosure of what a product will *not* do (the "restrictions apply" sort of thing), and with that in mind, I should provide the following caveat: This work is *not* intended to improve the way you deal with the nitty-gritty problems of daily life. I hope readers will, however, obtain an enhanced perspective about how that nitty-gritty fits into the larger scheme of things; also that readers will derive a good deal of pleasure from contemplating the very special ways in which our own kind experiences things—not to mention an appropriate degree of species modesty, given our natural tendency to set ourselves up as the Gold Standard by which 'Reality' is to be judged.

Although the worldview presented herein is strictly *Naturalist,* I recognize that more than half of serious modern readers frame their world within a context of *Supernaturalist* beliefs provided by our Religious Institutions. With that in mind, I will from time to time comment on a variety of traditionally held points that absolutely require updating if a person is to avoid the necessity of postulating *"two incompatible truths."* I'm referring to particular religious doctrines concerning human nature that are rooted in ancient times and that lack congruence with logic or modern science.

To promote simplicity in presentation, I've written the book in three overlapping divisions.

PART I emphasizes the fact that organisms can do no more than sample the information overflowing from their environment. Animals fashion the relatively scant information they're able to detect into representations of their surroundings—*approximations* that are good enough to promote survival and reproduction within a given species' ordinary world.

The information-processing equipment of our own grand and glorious species provides us with sophisticated representations of ourselves and our environment that far outstrip anything given to the other animals on planet earth. Nonetheless, the same principle ultimately applies. We detect only certain things, and we process the information we're able to obtain in only certain ways. Specifically, we've been naturally selected for adaptation to a world of mid-size-objects-and-their-movements. We cannot <u>not</u> experience the world in this manner because it's a core element of our information-processing programs. And though sophisticated sense-extenders (chemical assays, microscopes, radiation detectors, and on and on) have greatly increased the extent of information available, our processing systems still limit us to computations over *objects-and-their-movements;* and that continues to be the case even when using our conceptual equipment in its most advanced mode, as when we perform abstract mathematical manipulations.

PART II focuses on a crucial facet of our information-processing apparatus—one that's absolutely necessary if we're to respond adaptively to the complexities of our world—namely the need to choose what aspects of a situation need the most attention, and what sorts of responses are called for. The general solution to this kind of problem involves adaptive *weighting* of opposed behavioral programs, weighting that also varies under different circumstances. For instance, the urge to stop at a water bubbler will vary, depending on our degree of thirst. As in Part I, we'll make liberal use of introductory illustrations provided by less sophisticated organisms.

When dealing with the weighting apparatus apparent in our most sophisticated human systems, we employ the fancier word *'values'* rather than a more workaday term like *'weighting'*. The principle remains the same, however, so we might aptly say that our application of values can be traced back to weighting mechanisms that evolved in less complex organisms over preceding eons.

We'll focus on our most crucial value system, one that's absolutely necessary if an extraordinarily complex social species like *Homo Sapiens* is to make it through even a single reproductive cycle: namely our *sense-of-fairness* program. As with language, our sense of fairness begins to unfold spontaneously from an early age; and also as with language, one's particular 'language of fairness' derives in the first instance from that used by a person's familiars (albeit in more convoluted fashion than is the case with ordinary language).

What may be new to some readers is the notion that our human ethical sense (basically our *sense-of-fairness* program) can be accounted for within the context of Natural Selection. At a proximate level of explanation, no hypotheses beyond the ordinary workings of nature are required.

Part III focuses on the very core of our *human* existence, namely our consciousness. So central to us is this property that most of us wouldn't care to have our bodies kept alive if the areas of our brains underwriting this wondrous capacity were destroyed. For us to be *humanly alive,* that is, we need to register our world with full awareness, then be able to think and feel about it.

And though we're clear at this point in human history on the fact that a functioning brain is necessary for human consciousness, how this amazing attribute actually arises remains mysterious. For how can it be that a three pound cluster of corrugated flesh occasions such an astounding ability? Indeed it's almost weird, when you think about it, that the most immediate and most intimate aspect of our own selves remains so difficult to fathom.

We'll first review historical explanations given for human consciousness, most of them postulating the existence of a separable consciousness-substance. Problem with these traditional 'ghost' accounts, however, is that they fall into logical inconsistency when one takes the time to look at them more closely—which we will do.

After that, we'll address what we might refer to as the current *"Standard Model"* of consciousness. Only problem with this model, as we shall see, is that it doesn't work very well either. During our journey, we'll also see why some modern theorists of mind have been keen on getting rid of the whole notion of consciousness (or as their critics are wont to say, why such thinkers end up "feigning anesthesia").

When it comes to my own explanatory turn at bat, I plan to follow the strategy used by modern particle physicists. Whenever they come across something so basic that they can't break it down into more fundamental constituents, they simply accept it as a *given*. Force, for instance, is a given. The notion of force seems too fundamental to allow formal definition, and yet its effects on us are so compelling that to deny its existence in our ordinary world would amount to no more than intellectual posturing.

In similar fashion, consciousness also defies formal definition, because it can't be broken down into more fundamental constituents either. That being the case, I take the approach: *"Consciousness is another one of those fundamental items that's there all along."* The good news is that I don't even have to invent a name for such a theory, because there's been a label hanging around for centuries. Bad news is that the label, *panpsychism,* involves a notion that's also been pooh-poohed for centuries. We'll get to the reasons why in due course. But under the circumstances, I'm going to need some fancy footwork if readers are to take my conceptual model of consciousness seriously.

My editor, by the way, has asked me who would make up the readership for this volume. My optimistic answer was: *Everyone.* Or at least everyone *should* be interested in understanding more about the remarkable apparatus that provides us with our human understandings of the world, and everyone should be interested in getting a more detailed grasp on our limitations in comprehending *'What Is'*.

PART I
OUR HUMAN ILLUSIONS ABOUT THE WORLD

◆

(OUR HIGHLY RESTRICTED VIEW OF 'WHAT IS')

1

REALISM

We are all *Realists* in our daily life. That is, we take for granted our ability to see things as they are. Sure, we can be tricked on occasion by, say, visual illusions, but such illusions are correctable. We can also be tricked temporarily by false information, but once again this situation is potentially remediable. And, of course, our humanly attainable knowledge is always limited, so although we can know the truth about many things, those truths are necessarily incomplete.

The majority of people take the above tenets for granted, accepting such notions tacitly without any need to think them through. And if challenged as to whether things are *really* as they seem to us to be, they would unhesitatingly answer: "Of course, as long as we're talking about a normal person viewing things under ordinary conditions." And though this conviction is, as we shall see, clearly in error, the fascinating point is that the error has virtually no drawbacks when it comes to the day to day conduct of our lives. Nevertheless, since this belief does limit the potential richness of our human experience, there's much to be said for an investigation of how our particular species does in fact *represent* the world.

REPRESENTATION

I have italicized the word, represent, because some philosophers don't like the notion that we "merely represent" the objects and processes that go on around us. They're afraid that if all we have are human depictions of these items, then we're never going to be sure what's *really* there. They prefer to think, therefore, that we become "directly acquainted" with the objects around us, and that the complex sensory processes involved amount simply to mechanical steps along the way. But the fact that sensory detections provide no more than simplified approximations of our environment presents no practical problem at all, as long as these representations are good enough to get the job done—the job of adapting successfully to the *ordinary world* occupied by our species.

A USEFUL PERSPECTIVE

Sometimes we can get a better perspective on ourselves as individuals by looking at other people. In the present case, I think we can get a better perspective on our selves *as a species* by looking at how other species form representations that are good enough for successful adaptation within their own ordinary worlds.

The simpler the species, the easier it becomes for us to understand the sort of process that's going on, so let me illustrate the point by describing the behavior of the deerflies with whom I share my land. They make their appearance early each summer, circling my woodland gate posts like airplanes in a holding pattern. To accomplish this feat, not only does their visual <u>detection</u> apparatus need to register sufficient information, their <u>computational</u> systems have to process the information into a recognized pattern (let's refer to the pattern provided by a gate post as "vertical end-stopped object"). The pattern-recognition apparatus must then trigger appropriate programs housed in the deerfly's <u>motor</u> systems.

When I walk over to open one of my woodland gates in season, I can count on three things happening: As long as I continue to move, many of the deerflies will shift their attention to me. Their flight cycles will tighten, and they will move more rapidly. They will then attempt to land and insert their proboscises into my unwilling flesh. Notice that only two elements are necessary in order to evoke this feeding effort on their part: a vertical end-stopped object (in this case, myself) that is moving.

Note in turn what an effective <u>engineering shortcut</u> this amounts to, because almost all vertical end-stopped objects moving within the *ordinary world* to which the deerfly has adapted will consist of midsize animals (*aka* lunchtime opportunities). And we need to emphasize: such engineering shortcuts are crucial for an organism of this size, given the diminutive brain space available for its *computational activity*.

Let me illustrate the importance of computational space by providing an analogy from my personal computer activity. When I purchased my original Macintosh, the computer's central processor and its small black-and-white screen fitted into the same compact housing, keyboard in front, small dot-matrix printer to the side. With the passage of years, I now have a much fancier setup—large color monitor, expanded keyboard, freestanding hard-drive, high-speed modem, scanner, laser printer for first-rate text printouts, and an additional printer specialized for color reproduction. My small study has become overwhelmed, so to speak, with my computational equipment, and there's no space left for any additional stuff. Since a deerfly's computational space is hardly larger than a grain of sand, it's not surprising that its own computational space gets used up pretty quickly.

And my easy reference just now to "computational activity" indicates how far our familiarity with the sort of apparatus underlying animal behavior has come. Not too many years ago, the workings involved in an insect's ability to perform rapid and complex behaviors seemed almost totally mysterious. But in recent times, at least two areas of scientific research have greatly demystified the situation. First, advances in neurophysiology have helped us to understand the way particular elements of the nervous system detect and process useful information from the environment. Secondly, computer technology has provided many helpful analogies by which to make sense of the nervous system's activity. When one's personal computer performs this or that miracle after being loaded with the proper program, it's easier for us to become comfortable with the notion of 'programs' bundled into the nervous systems of organisms. So if we apply a computer analogy to the deerfly's feeding behavior—albeit in the form of an algorithm that's miles removed from actual machine language—we might write: compute whether the information detected fits with the pattern of a vertical end-stopped object > if so, compute whether the object so identified is in motion > if so, connect with output programs to tighten flight spirals in the direction of the object > once within close approximation, initiate additional output programs, first for landing, and then for feeding.

This example illustrates that relatively simple programs, using a modest amount of input information, can underwrite amazingly complex adaptive maneuvers. However, all such *engineering shortcuts* exact a price. We can see this easily by observing additional deerfly behavior. When I drive my Toyota pickup through my woodland during the summer, say, to gather a load of firewood that I've cut and split earlier, deerflies swarm around my moving vehicle. Many of them will manage to land and then attempt to insert their proboscises into my truck's unreceptive metal. Why are these critters wasting their precious time and their limited energy in such a quixotic endeavor? Because their detection systems are registering no more than "vertical end-stopped object in motion," and that's enough to trigger their approach and feeding programs.

Needless to say, pickup trucks did not form part of the *ordinary world* in which the deerfly originally evolved. And when these insects are confronted with situations that call for cognitive skills beyond those naturally selected for coping with their native woodland surroundings, their systems fall into error that is both systematic and predictable. By contrast, I can readily distinguish a white tail deer from a Toyota pickup—and I can do it effortlessly, even on my worst day.

SOME ANIMALS FORM FULLER REPRESENTATIONS THAN OTHERS

Some other animals, it must be said, can also perform this feat of detection. My dog, Shep, for instance, will bark at a deer, but will come running with tail wagging when I drive up in my truck. And when I get out, he easily distinguishes me from my Toyota, jumping enthusiastically into my arms so we can renew our bonding at closer quarters. And he can just as readily distinguish sheep from deer. In the former case, he never barks; instead, he approaches slowly, eyeing my sheep commandingly, a behavioral program that's built into border collies.

And though, granted, none of the other animal species see things as they *really* are, some animals represent objects and the relationship between objects more fully than others. For instance, when Shep was a pup, he had not the slightest difficulty in recognizing his new food bowl. And when I moved the bowl from its usual position in front of the kitchen fireplace out to the meadow immediately behind my cottage, he had no difficulty in distinguishing his familiar bowl from the ground on which it was set. But then I gave him a more difficult problem: While he was in the meadow down back, I placed his bowl on the near side of the sheep fence. He came dashing right up (maybe figuring like ourselves that a straight line is the shortest distance between two points), but was unable to get to his food. I then walked along to the nearest gate about a hundred feet away, and helped him find his way around and back. One experience was all it took. From that time on, he employed his new-found appreciation of relationships among objects to move from one fenced area to another by circling to whatever gate was nearest. He also had no trouble in *generalizing* from his original motivation (food), working his way from one fenced area to another in order to shepherd a wayward ewe, or for the pleasure he takes in my company.

My sheep, it must be said, have also been capable of learning to move from one fenced meadow to another by using open gates. But it took them much longer to get the hang of it, and their comprehension of one gate-setup did not generalize very well to other fenced regions with different land markings. So while dogs and sheep both form useful representations of what *really* is, dogs do so more adeptly, which of course is why we judge that dogs are smarter than sheep.

REPRESENTING THE *REAL* SELF

Sheep, along with almost all the other animals, have such a limited representation of what *really* is that they can't even recognize themselves—and I can supply a personal vignette to illustrate the point: During summer heat waves, I have a habit of retiring to an area I've set up in my naturally 'air-conditioned' cellar. One day while I was basking in cool there, my reading was interrupted by a loud crash. As I looked up from my book, I saw glass flying everywhere from the cellar window above me, and *mirabile dictu,* one of my dominant ewes was standing on the floor right in front of me. She hadn't even cut herself and, remarkably, had even managed to land upright after falling six feet through the window.

As it turned out, the late afternoon sun had been glinting off the window in mirrored fashion, and as she was grazing nearby, she saw a ewe who failed to give ground at her approach. Taking further umbrage at this ewe who was daring to react so assertively, she charged—right through the window—in order to put the inferior creature in its place.

I have to reluctantly admit that Shep fares no better when it comes to perceiving the reality of *him*self. While he spots members of his canine species with a facility that sometimes amazes me, he shows no evidence of being able to identify himself. If we look in a full length mirror together, for example, I see the *reality* of me with no difficulty. But Shep seems to see only me and some dog—his relatively sparse representations do not even afford him the opportunity of recognizing himself. Of the other animal species, only chimps seem to definitely have the computational 'meta' systems necessary for such self-recognition.

IS *HOMO SAPIENS* SIMILARLY LIMITED?

Since all the other animal species on our planet elaborate no more than relatively sparse representations of what is, it's likely that even the species topping the animal feeding chain will be limited in kind. After all, there does seem to be a sliding scale of representational complexity—and for all we know, that scale may place dogs even further above deerflies than we are above dogs. So if that's the case, we would expect that our *representations-of-what-is* may be more complex than the others, yet for all that involve only shadows of what's really there.

To society's Guardians of the conviction that we can indeed know what *really* is, such a judgment seems difficult to accept, and they sometimes charge that the sort of account just given leads to nothing more than a bankrupt know-nothingism. So with that concern in mind, let's look next at what we might call the problem of *skepticism.*

2

ON BEING SKEPTICAL

Just as we are all *Realists*, we are also all *Skeptics*—to a point. If for instance a man asks a woman for a date Saturday evening, and she answers "Sorry, I'm busy," he's probably a bit skeptical about the literal truth of her response. And in the public arena, we may frequently be skeptical about the wondrous effects that a politician assures us will follow from his new tax plan or foreign policy. Furthermore, if we see something that's unusual to the extreme, we may even wonder if we're seeing things accurately, double-checking with a companion perhaps by asking: "Hey, do you see what I see?"

What, however, if our perceptual apparatus were inherently flawed? Then no amount of additional flawed-information would be corrective because, as the computer gurus put it, "Garbage in, garbage out." No amount of asking "Hey, do you see what I see?" will help the situation if all our companions have the same inherently flawed perceptual apparatus. But if that turned out to be the case, should we not then quite properly doubt *everything*? Such a prospect greatly troubles Guardians of the conviction that we can indeed know what really is. They feel relieved, therefore, that the position of doubting everything—called *Universal Skepticism*—won't pass muster.

Here's how my old college textbook in Scholastic Philosophy expressed the point: *"Universal Skepticism is theoretically absurd and practically impossible."* Universal Skepticism is theoretically absurd because if a man were to seriously claim that "Nothing is certain," then such an affirmation would have to apply also to his own firm contention that "Nothing is certain." He would in effect be contradicting himself. Furthermore, Universal Skepticism is practically impossible because we in fact live our daily lives with the operating conviction that we need to do certain things in order to maintain our lives, our health, and our quest for happiness. If the *apparent* car coming down the street were no more than an arguable possibility, we would not treat it with the firm degree of caution that we so wisely exhibit.

CONTROLLED SKEPTICISM

But between Universal Knowledge and Universal Skepticism, there are countless intermediate positions. Scientists, for instance, speak of building models of reality—a *model* referring to the relatively simple representation of some actual large-as-life reality. The goal of theoretical model-building is to provide constructs that are humanly useful, *useful* as tools that enable us to inch (or occasionally jump) forward in our comprehension of what is, and tools that sometimes make possible immediately helpful applications.

Let me illustrate: When I was a medical student in the mid nineteen fifties, standard treatment for high blood pressure consisted of small doses of phenobarbital—a relaxing sedative that was of pathetically little value in the control of significant hypertension. A promising agent introduced around the time consisted of extracts from *Rawulfia Serpentina*, a plant that had been used medicinally in India for a variety of problems since ancient times. The 'new' medicine produced a gratifying decrease in blood pressure, but unfortunately was all too frequently associated with serious psychiatric depression as a side effect. Scientists, who had been studying signaling chemicals in the brain, discovered that *Rawulfia* extracts depleted the brain of some of these chemicals (called *monoamines*).

Around the same time, great strides were being made in the medicinal treatment of tuberculosis—enough that most of the tuberculosis sanitariums in the Country were soon able to be closed. One of the new antibiotics was a chemical that inhibited the body's enzyme for breaking down monoamines (hence, the name 'monoamine oxidase inhibitor' or simply MAO inhibitor). While the new agent had serious side effects and turned out to be of limited value as an antibiotic, doctors noticed that many depressed sanitarium residents who'd been given the medicine became free of their melancholia. Following this serendipitous observation, newer and safer MAO inhibitors were developed, and they became the first actually efficacious anti depressant medicines.

Now it was time for scientists to provide an *explanatory story* for what was going on, based on these two pivotal facts: (1) a chemical that depleted the brain of monoamines often led to depression, (2) agents that interfered with the breakdown of monoamines, thus allowing these chemicals to hang around longer, often succeeded as antidepressants. So perhaps the long sought biological underpinning of depression had finally been discovered. To wit, monoamine signaling chemicals in the brain are necessary to support a normal mood. Deprive the brain of enough of these necessary chemicals, and the unfortunate person will fall into depression. Restore these specific neurotransmitters to their normal range, and the patient's depression will go into remission. This is the so-called *Monoamine Theory of Depression*, a conceptual construct that has led to an amazing amount of fruitful research during the years since.

As one might expect though—or at least as scientists have come to routinely expect—this initial model turned out to be no more than a stick-figure oversimplification of the pertinent brain mechanisms that were at work. The reality turned out to be far more complex, and scientists currently consider that they're barely through the doorway to understanding the fullness of what's involved. But, one might ask, have we not at least achieved some preliminary albeit *partial truths* in this area? Well, yes and no, a notion that we'll come back to later; but for now, let's return from the level of our conceptual knowledge to the more basic level of the *startup information* with which we work. Specifically: can we trust the evidence of our senses?

DO OUR EYES DECEIVE US?

We humans (as well as the other primates) have a highly developed sense of vision. We see individual objects sharply, in three dimensions, and in full color. It's fitting therefore that we focus first on this honcho of our detection systems to see if its outpourings are trustworthy. Before getting into specifics, however, let me underline a point I made earlier about the other species on our planet: <u>Animals need not perceive 'what is' as it really is; all they need do is construct representations that are good enough for successful adaptation to the ecological niche in which they've evolved</u>.

With that general notion in mind, let's examine the sort of information that our visual system gathers. Note first that it detects only one kind of energy, electromagnetic radiation (EMR), and that it samples only a minuscule portion of the total EMR available. To be specific, our retinal detectors respond only to wavelengths between 430 to 700 billionths of a meter, representing about 1/300,000,000,000,000,000,000,000,000,000,000,000th of the complete spectrum! We're obviously missing a bit of what's going on. How come?

Once again, as with the deerfly, the answer is to be found in the notion of engineering shortcuts. Receptor space in our eyes is limited. Hence, we are forced to *sample* the ambient EMR. But why not, one might ask, take smaller samples of wider girth, sweeping across a larger portion of the complete spectrum? Well, by happy coincidence, it turns out that our receptors sample from just those wave lengths that our very own sun is most fond of emitting—not really a coincidence at all, of course, since evolving animals whose visual detectors were more congruent with the sun's output had an adaptive advantage on our planet, and so were naturally selected.

But one might ask: has our visual system not provided access to *accurate*—though granted incomplete—perceptual knowledge? Once again, the answer is "Well, yes and no." Let me illustrate: Butterfly aficionados have long admired the *Eurema Lisa*, a yellow winged creature of astonishing beauty, whose male and female versions are visually indistinguishable. But how is it (other than through the mystery of love) that males and females of this species can readily tell each other apart at twenty paces? Turns out that their visual detectors react to light in the (invisible to us) ultraviolet range. The upshot is that they see stripes on a male's wings, while we see solid yellow. One might be tempted to ask then: "Well, what color are the male Eurema's wings *really*? Are they really yellow or are they really yellow with stripes?"

Before addressing the question directly, let me tell you a story: Seems a cowboy was on his night out. This woman sits down beside him at the bar and asks "You a cowboy?" He doffs his ten-gallon hat and says "Well, I dress like one, I work on a ranch, I ride a horse, I herd cattle, and I mend fences, so I guess you could say I am one."

"Well," she says, "I'm a lesbian. I get up each morning and start thinking about women. And when I go downtown, all I do is look at the women walking by. And when I get home, all my daydreams are about women. That's all that seems to be on my mind." A while later, a nice young couple sits on his other side, and they also ask "Are you a cowboy?" to which he replies: "Gosh, I always thought so, but I just found out I'm a lesbian."

Now if we were counseling that cowboy, we might point out that sometimes in order to be sensible you have to take into account not only the data but the specific organism that's involved. Same with the earlier question about the Eurema's 'real' color. We have to take into account not only the perceived object but also the perceiving organism. That is to say, the act of perception always involves a *relationship*. It's true to say that "the *Eurema Lisa* is a yellow-winged butterfly," as long as we're prepared to add the qualification, "as seen by a member of *Homo Sapiens* with a normal visual system under conditions of ordinary light." Naturally, we don't usually bother to add such qualifications, because the way something *dependably* appears to members of our species is almost always what we're interested in. Stated another way, as long as we can count on our visual representations being stable, and as long as we can associate a given visual representation with dependable properties (e.g. flames are hot), we're in business.

So for those who want to insist that our eyes should let us see things as they *really* are, we must answer: "Our eyes do deceive us." But if we ask a different question, "Do our eyes provide us with <u>humanly useful</u> representations of our surroundings?" then the answer is "Yes they do, and in this sense our eyes do not ordinarily deceive us." Put another way, vision has not evolved for the benefit of a few philosophers who want the comfort of knowing 'what is' as it *really* is. Our visual mechanisms have been naturally selected because they give us a quick and useful read of environmental features that are crucial for adaptation to our ordinary world.

To hammer home the importance of a species' *ordinary world*, notice that the way our own visual systems happen to represent the Eurema Lisa's wings is of little practical import to us, but it's crucial in the ordinary world of that species to identify sex differences. Otherwise, their ability to reproduce would be seriously compromised.

ANOTHER RELATIONAL ASPECT OF OUR VISUAL SYSTEM

In addition to colors that appear different in relation to different species, colors also look different within our own species in relation to other aspects of what's being viewed. To convince yourself of this fact, take a look at your TV while it's turned off. Most picture tubes will appear gray. Now turn the set on and switch to one of your 102 cable channels—the one that's rerunning an old detective show right now—and wait until the sleuth gets to hunting around in the dark with flashlight in hand. At this point, you will probably see a bit of his face, along with the bright beam from his flashlight, while the rest of the scene will be pitch black as befits a night scene.

Now try to figure out how the TV tube manages to *project* the color black. After all, we learn in school that black is not a color but the absence of color. So how does one project absence-of-color onto the gray screen? It's not as if the tube is up to spraying black paint—pigment that absorbs all wavelengths of visible light so they're not reflected—and then gets to rubbing paint remover across the screen for the daylight scene that comes next. Turns out the answer is to be found, not in the TV set but in our own visual system, which has deepened the contrast between the detective's flashlight beam and its surround.

And even this account considerably oversimplifies the situation, as may be seen from this test question given to students in visual physiology: Which reflects more light, a black card in bright sunlight, or a white card in the shade? "Well," an intelligent student might answer, "I know that black absorbs all the colors—that's why it looks black—while white reflects all the colors—that's why it looks white—so the white card reflects more light, even though it's in the shade." Wrong! The black card reflects more light under these circumstances. Then why does the black card look darker than the white card? The answer, my dear Brutus, is to be found, not in the cards but in the adjustments made by our own visual apparatus.

The computational circuits of our visual systems make many such adjustments in what we see. Once again, of course, as with all simplifying engineering programs, there's a price to be paid. For just as some philosophy buffs would have preferred that nature provide us with visual systems enabling us to see objects as they "really are," some photography buffs would have preferred nature to provide our eyes with light-meter capabilities. But alas, our visual systems are terrible at the task of measuring light intensity, so engineers have had to construct devices that let the photographer know the *real* amount of light that's bathing a scene.

We should hasten to add, however, that Nature has not been unkind here. Once again, our visual systems have been selected on the basis of adaptive efficiency, and they provide something more important to us than objective measurement of EMR intensity, namely, relatively stable representations of objects despite rather marked changes of light intensity. So effective, in fact, is our system for making adjustments that my red pickup will still appear the same color to me whether it's parked in a sunny meadow or in woodland shade—even though the mix and amplitude of electromagnetic radiation reflecting from its surface is meantime varying all over the place.

This color constancy is part of our apparatus for maintaining object constancy, and object constancy is crucial to us for successful adaptation within our species' ordinary world of mid-size-objects-and-their-movements. So let us now examine object constancy in more detail.

3

ABOUT OBJECT CONSTANCY

A traditional philosophical chestnut goes like this: Suppose a tree falls in a forest with no one around. Does it make any *sound?* Before analyzing the import of this old question, let me tell you a story:

A young boy visits his grandfather for an overnight stay. First thing they do is have lunch together, but the boy's a bit put off by the fact that the plates aren't all that clean. When he complains, his grandfather answers "Well, they're as clean as coldwater can get 'em." Same thing happens at dinner, and the grandfather gives the same explanation, "They're as clean as coldwater can get 'em." Next morning, they have pancakes and syrup, because the grandson says that's his favorite breakfast. He's unable to finish it, however, although by now he doesn't even bother to complain about the condition of the plates. He's just glad it's about time to go home. But when he gets to the front door, his grandfather's big ole dog jumps up on him, almost knocking the boy over. "Grandad," he complains, "will you tell that dog to leave me alone," which gets the old man to admonish his pet severely: "Coldwater, *get down!*"

I'm not sure if the boy felt any better about the dishes at that point—I know his mother didn't when he told her what had happened—but in general, it *is* helpful to know what a person's referring to when he uses a particular word. The same thing applies, of course, when a person refers to 'sound', which by coincidence also refers to two different things. The word can apply to the BAM! that I hear after I've chain-sawed a tree to the ground for next year's fire wood. Or by extension, I can use 'sound' to refer to the rapid oscillation of air molecules set in motion as the falling trunk hammers the forest floor. In ordinary speech, we often mix these two meanings together, referring either to the sensation of sound, or to what has caused the sensation (or to both). But once these different uses of the word are untangled, we have no problem at all in answering the old chestnut.

We would say: "If you're referring to the subjective experience of sound, then if no one's listening, there's *no* sound when the tree falls. If you mean by sound an adequate stimulus that *would* produce the experience of sound if a person with normal sensory apparatus were present, then the falling tree makes one helluva' sound."

We do that same running together of subjective experience and stimulus-apt-for-evoking-the-experience when we're talking about *any* of our senses. "Sweet as maple syrup" describes a substance as sweet because it can evoke a sensation of sweetness in us; "Spicy perfume" refers to a volatile chemical that ordinarily evokes that kind of sensation; we say "smooth as silk" when describing a garment that evokes the sensation we'd get from running our hand across a fine piece of silk; and "white paint" denotes a pigment that will reflect the full spectrum of visible light, evoking in us the sensation of white.

It's no accident, however, that plain speaking people don't bother much with the above distinction. They're totally occupied with the human reality—the reality of our ordinary world as it appears *dependably* to us. It's only when we get to fancy thinking that problems start to arise, especially for Guardians of the conviction that we can indeed know what really is. They're afraid that if our sensations tell as much (or more) about ourselves as about the things around us, then how are we to be sure we know these things as they *really* are. And the old philosophical chestnut about the falling tree was sometimes used also as a stand in for this more general question: "Do *any* of the things of our ordinary world exist independently of our perceptions of them?"

PRIMARY AND SECONDARY QUALITIES

Some philosophers of yore, we should note, were quite aware that sense experiences occurred as the result of interaction between an external stimulus and the characteristic detector systems of a given species. They knew, for instance, that dogs detect sounds too shrill for the human ear to hear. As for the relativity of sensation *within* the same species, they could call on the familiar fact that tepid water feels warm to a hand that's just been removed from coldwater (not the dog), and that an odor gradually fades with continual exposure to an object that's causing the sensation.

But their understanding that our perceptions tell at least as much about ourselves as about what's outside ourselves troubled Guardians of the conviction that we can indeed know what *really* is. So they would make a distinction between, say, the taste of a block of salt in my barn and other properties of the block such as its size and shape. The great John Locke, for instance, would have called the block's salty taste a "secondary quality" because that quality does not exist <u>in</u> the block itself. But he would refer to the block's size and shape as "primary qualities," based on the fact that a salt block of certain dimensions really is setting there in the barn. The taste may be present only when some animal is licking the salt, but the block is there *objectively*, that is, independently of whether anyone is there to affirm its existence.

Only an unusual maverick would dispute such an "evident truth." But fortunately—for our entertainment—a famous theorist of yore named George Berkeley argued just that opinion, and he did so quite eloquently. His famous dictum went *"To be is to be perceived"* (being a Churchman, the "to be perceived" part meant ultimately to be perceived by God). However, I said "for our entertainment," because even if some of Bishop Berkeley's readers don't believe there's a *logical* flaw in his reasoning, they still most often react to his arguments as they would to a magic show: "Wow, this guy is amazing; I don't know how he manages to pull off this trick…but of course I know it isn't really so."

And we *know* it isn't really so because our pre bundled 'reality program' is inexorable. Even when this program breaks down badly, as when psychiatric patients suffer from conditions involving "feelings of unreality," they will *still* take the actual surroundings of their ordinary world into account—unless their acute conditions are further exacerbated by compelling delusions and hallucinations. Except under the most extreme set of conditions, that is, we're simply not able to functionally disbelieve in the existence of the objects around us. Nature has seen to this by embedding our 'self-preservation programs' so deeply. Put another way, individuals lacking the fundamental 'sense-of-reality' program would promptly select themselves out of existence. Of course, thanks to the wonders of human language, academicians who care to do so can entertain themselves by word-playing such disbelief to their heart's content.

Note by the way, as Bishop Berkley himself was happy to point out, the fondly embraced distinction between primary and secondary qualities tends to break down anyway, because knowledge of the so-called primary qualities of an object depends on our sensing its secondary qualities. The reason we're absolutely convinced that material bodies of given extension are *really* there is not then that we see primary qualities in some mysteriously superior way, but that our systems will not allow us to detect things except as objects and their motions.

A WORLD WITHOUT OBJECTS?

At the end of last chapter, I noted that <u>object constancy</u> is crucial for successful adaptation within the *ordinary world* of our own species. And I went on to characterize that world as one of mid-size-objects-and-their-movements. In order to emphasize the robustness of this human world of ours *to us humans,* I will ask readers to conduct the following experiment: Try to imagine a science-fiction world devoid of objects. I think you will find this project a most difficult undertaking. And I might add, even if we could succeed in performing such an exercise of the imagination, we certainly would have no reason to *want* to exist in such a world, because all the people and all the things we love would be absent (our own selves included).

VISUAL OBJECTS

Focusing again now on our *ordinary world,* let's use vision once more to illustrate how this system goes about the task of constructing midsize objects and their movements. The computational process starts right away, even before the information detected leaves the eye on its journey via the optic nerve to the brain.

But before outlining the process that's involved, let me seize the opportunity to review a point made earlier, namely, that scientists think of themselves as constructing *models* of reality, and that eventually a given model will be replaced by a more useful one—never that scientists will have finally learned *"the truth, the whole truth, and nothing but the truth"* about whatever it is they're studying. Some of the most productive models have consisted of analogies drawn from better understood mechanisms. When fashioning a conceptual model of the eye, for instance, it was natural to think of a camera. After all, cameras also "picture things," and since these devices had been engineered by us humans, we knew what we were talking about.

Here's how science textbooks used to outline the model: Just as the camera has a lens in front to focus an image on the film placed along the camera's back wall, so the human eye has a lens that focuses light on the retina attached to *its* back wall. And just as the camera can change the width of the lens opening to let in more or less light, the human iris (brown or blue) varies its size to let in more or less light. But scientists have had to alter that part of the model in which the retina was viewed as equivalent to a film emulsion. That's because we now know there's a crucial difference between the relatively passive registration of light by the film's silver-salt emulsion and the intensely active information-processing that's going on in the retina.

To construct a useful picture of what's happening, think of a circular target composed of detector cells (e.g. *cones*), all sending their output to the same retinal *ganglion cell*. When this cell receives adequate stimulation, it will fire off a message to the brain. One would think that the more detector cells activated, the greater the ganglion cell stimulation would be, and hence the more vigorously it would send messages to the brain. Puzzlingly—at first—this turned out not to be the case. In fact, it was as if the more widely a light stimulus was applied, the *less* strongly an affected ganglion cell would send out its messages.

Eventually, scientists figured out that a circular target of detector cells connecting with the same ganglion cell produces different effects. Activated cells within the target's bulls-eye *stimulate* the 'on center' ganglion cell. But cells in the surrounding target area actually *inhibit* the ganglion cell, tending to negate the stimulatory effect coming from the bulls-eye. Given this sort of arrangement, the most potent stimulus consists of bright light falling on the central bulls-eye, with much less light falling on the surrounding target area. This condition, of course, produces *maximum* stimulation and *minimum* inhibition.

Turns out that this sort of stimulatory package is found most readily at object *edges*. To provide yourself with evidence of the fact, look around the room where you're sitting now, and you'll probably notice how one side of an edge looks brighter than the other, the brighter side of the edge dependent on the direction of the lighting source. The ganglion cells we've just described are ideally suited for registering the various edges of the objects that surround us.

When this information reaches the visual cortex, certain cells will respond strongly to stimulation that's been relayed from groups of these ganglion cells. Now, however, the preferred stimulus consists of a straight-line series of activated ganglion cells running at a given angle. That's why these particular neurons in the visual cortex are called 'edge cells', because in effect they detect or construct *edge lines* from the field of cells they're responding to.

This information is in turn relayed to more anterior cortical areas that use intersecting edges to construct the discrete objects of our visual experience. And if the line of stimulation moves as a unit along a course of edge cells responding to the same edge angle, the "same object" will be detected in motion. Object-and-motion computations, it should be said, are performed in different (albeit closely connected) brain areas, the processes proceeding at the same time with mutual feedback—hence the name, 'parallel processing', referring to an approach that's far more complicated than the one-step-at-a-time approach ('serial processing') used by our personal computers.

ARE OBJECTS DELINEATED OR CONSTRUCTED?

To summarize, on-center ganglion cells in the retina respond best to sharply focused light with a duller surround, the sort of conditions found most commonly at edges. When a series of ganglion cells ranged along a given line become activated together, this stimulus is selectively picked up by 'edge cells' in the visual cortex. Additional cells from higher brain areas then use this information to construct the multi edged objects of our ordinary experience. Whenever 'edge' stimulations move in concert across brain areas assigned to this task, the same coherent object is represented as in motion.

Here's how most neuroscientists apply their common sense when thinking about this process: Discrete objects exist in our surroundings. Light bounces off these objects, producing a differential light pattern, especially at the edges of these objects. Our visual system then sharpens these edges, producing the crisp delineation of objects that we find so useful in dealing with the world.

Here, however, is where a skeptic might wonder: Since the visual system amplifies the differences in local light intensity, especially when these differences are aligned along one or another angle, perhaps our visual system is actually constructing edges and in effect *creating* the discrete objects of our experience—not just delineating discrete objects that are actually there. In the upcoming chapter, we will focus on this issue.

4

OBJECTS OF OUR EXPERIENCE

At first glance, it seems so utterly easy to quash a skeptic's lack of common sense when he wonders whether the physical objects of our daily experience actually exist that we might for the moment humor him by considering a thought experiment—more than merely an idle exercise of our imaginations, however; rather a seriously intended model of reality proposed by a highly regarded theorist. His name was Albert Einstein, and here's how he thought about the issue:

> Could we not reject the concept of matter and build a pure field physics? What impresses our senses as matter is really a great concentration of energy into a comparatively small space. We could regard matter as the regions in space where the field is extremely strong. (242)[1]

Einstein, it must be said, was never able to work out the detailed equations that would be required to consolidate this part of his field theory, yet given the transposability of matter/energy, his thesis does not simply invalidate itself. So let's for the moment see what might follow in the day to day conduct of our lives if we were to accept such a notion.

Interestingly, *nothing!* That is, nothing different would follow in our daily life—except perhaps a greater appreciation for the elegant engineering shortcuts adaptively employed by our nervous systems. Try to imagine, for instance, a world in which there is but one all-encompassing field of force ranging vastly in concentration. Then consider the computational savings for the human brain (given its limited computational space) if this ever varying continuum could be chunked into manageable blocks.

The advantages provided by such an approach should be easy for us to appreciate, because we're already familiar with the use of *bundling* strategies at a level of daily life to which we have conscious access. Suppose, for instance, that we meet a new friend in our neighborhood who provides us with her telephone number. The area code (3 digits), local exchange (3 digits), and particular number (4 digits), come to a total of 10 items, exceeding the 7 or 8 item maximum provided by our immediate-memory function. Yet we have no problem, because we bundle the three digit area code into one familiar unit and deal similarly with the three digit local exchange. So all we really have to concentrate on are the four digits of our friend's particular number. Or think of the number bunching allowed by the IRS, enabling us to 'round off' each item to the nearest dollar. Sure makes our monetary manipulations easier at tax time. In similar fashion, "rounding off" field fluctuations into discrete objects would likewise help greatly, even if this particular bundling occurred so automatically that we were never conscious of its occurrence.

CORROBORATING EVIDENCE CONCERNING THE REALITY OF OBJECTS

Let's return now from this imaginative foray back to our ordinary world. We had left off with the skeptic wondering whether the way our visual system operates wouldn't *create* discrete objects, even if such objects did not actually exist. In response, we might well have pointed out that the representations provided by our visual system are strongly corroborated by our sense of *touch*. I not only *see* that my truck's body ends at its tailgate, I can feel along my pickup's smooth surface, and my hand will eventually curve around at the truck's exact visual endpoint. From there, I can trace at right angles across the tailgate to the other side. And I might add, as I'm pacing off these dimensions, there's no way I can walk *through* my vehicle.

THE QUANTUM *LEVEL* OF DETECTION

Yet the determined skeptic might still argue with us, pointing out that the individuality of my pickup—its forming one discrete object—depends on my visual system's particular *level of representation*. Suppose, for instance, that my eyes detected things at a resolution employed by some of our modern sense-extenders. Then I would visualize things quite differently: Each metallic atom that helped to constitute my tailgate would be distinct from its surrounding atoms, so the gate would appear as a manifold composed of countless tiny pieces, each at a distance from its fellows. At this exceedingly microscopic level of visual detection, in fact, it would be hard to know that the tailgate was indeed one piece until testing showed that its plethora of separate components all moved synchronously. Even then, we might conclude that the tailgate's unity involves more the functional unity of a system than the oneness of a single discrete physical object.

Progressing even further into this microscopic world: If we could view the scene *within* each of the component atoms, we would at this minuscule level detect a number of electrons moving around the atom's periphery at (relatively) huge distances from its central nucleus. Not surprisingly under these circumstances, when scientists first attempted to picture the atom's innards, they started by imagining a planetary model. That is, just as the planets of our own solar system circle the sun, individual electrons were to circle their nucleus. But this model broke down in a number of ways, highlighted by the experimental finding that electrons just don't act like participants in our ordinary world—which reminds me of a story.

Seems this wayward wife was entertaining in her bedroom when her husband unexpectedly arrived home. "Quick," she said to her new friend, "hide in the closet," which he proceeded to do. But her husband immediately became suspicious, so he started looking under the bed and behind the curtains. Finally he opened the closet door and found the man huddled in a corner. "What are you doing there?" the husband demanded angrily, to which the man replied—after a somewhat awkward pause—*"Well, everyone has to be somewhere."*

I'm told the husband was not at all reassured by the explanation, though as far as it went, the fellow's response was perfectly in line with a principle that applies unfailingly in our ordinary world. That is, everything has to be located in some *definite spatial position.* But unfortunately, that notion doesn't fit experimental results at the *quantum level* of reality. So although physicists continue to call the electron a 'particle', the electron is *not* like, say, a teensy dust particle within our ordinary world. That's because the electron also behaves like a wave, and waves of their nature spread out instead of occupying discrete positions in space.

So the skeptic would no doubt end up noting that the individuated objects of our daily experience depend on the *mid level* detection and analysis that our systems use. Not only does that apply to our visual system, but also to our sense of touch. The totally smooth and continuous surface of my pickup—as reported by my hand—would be pocked by a series of gigantic chasms from the perspective of tunneling microscopy.

Note, however, that the actual level of detection provided by vision and touch works superbly within our *ordinary world.* To emphasize this fact, imagine a scenario in which human eyes revealed their surroundings at the level provided by an electron microscope. We would never be able to respond to the midsize aggregations of our ordinary world (e.g. cars) in the timely fashion needed to avoid being run over. And if our sense of touch presented us with the humungous amount of data per unit of space that's provided by tunneling microscopy, our computational resources would be totally overwhelmed before we could move our hands across a finger's breadth of surface. In regard to both vision and touch then, the <u>level of representation</u> provided by our detection systems enables us to deal optimally with our *ordinary world* of midsize objects and their movements.

PERCEPTUAL OBJECTS > CONCEPTUAL OBJECTS

For many millions of years before *Homo Sapiens* appeared on the scene, increasingly sophisticated animals were honing their sensory systems. More accurately stated, animals with more effective systems for detection had a leg up when it came to survival; hence they and their progeny were naturally selected in an ongoing cascade of replication. Our own species rose to the top because increasingly effective *symbolic systems* gave us a huge additional advantage. That said, however, our language system came on the scene long after *discrete objects* had already been selected by higher animals as the perceptual coin of the realm. So we will need to examine how this fact shaped the way in which that gloriously complex symbolic apparatus of ours developed. But let's set the stage for that investigation by first focusing on the notion of symbolism itself.

5

SYMBOLISM

A symbol is *anything that stands for something else*. That being the case, a symbol doesn't become a symbol until someone *uses* it that way. If a person lights one lantern in a church steeple just for the heck of it, we might not think of it as a symbol. But if he said to Paul Revere beforehand, "Okay, one if by land, two if by sea," *that* particular lantern display would definitely constitute a symbol.

Though members of our grand and glorious species use symbols to a farethee-well, it's not as if we're the only ones to do so. For instance, whenever I walk through my woodland, a series of sentinel crows will tell the other crows about my presence. Since the loud CAW stands for something in the direction of "Look out fellows," the CAW is not just a noise, it's a *symbol*—even if the other crows aren't up to calling it by name.

Shep uses a bunch of symbols when he's communicating with me. For instance, when it's time to put the sheep in (an activity that's right down his alley), he'll come over to the chair where I'm likely to be reading and look at me expectantly. If I'm slow to respond, he'll wag his tail. And now that I think of it, his body will be doing a bit of waggle dancing too—all of that activity being *about* something else, something in the direction of "Let's get going, we've got work to do." And when I nod, he starts moving toward the door right away. That's because he understands my bouncing head is not just a bouncing head; it's my way of affirming "Okay, so let's go."

ANIMAL 'LANGUAGE'

Vervet monkeys use acoustic signals in a much fancier way than the crows on my homestead. I've been told that Vervets have three highly distinct alarm calls. If they use one, their fellows will start climbing the nearest tree; if a different one, they'll duck beneath the surrounding brush; and if a third, they'll stand up straight and start scanning the ground. No coincidence, because each call correlates with a specific danger: Call number one with a leopard (no way that large feline can make it out to a tree's smaller branches), call number two with a hawk (hard to scoop a monkey on the fly when it's protected by a tangle of low lying branches), and call number three with a snake.

While empathy for our fellow primates might make us want to glorify this 'language' use by Vervets, the lowly honey bee seems even further ahead when it comes to the use of symbols. That's where the famous waggle dance comes in. After spotting a source for making honey, scout bees will return to their hive and go through a characteristic series of motions that convey information *about* the source—its direction, its distance, and its richness.

HOW WE USE THE WORD, LANGUAGE

Despite impressive use of symbolism by such species, however, I've had to put the word *language* in scare quotes in the preceding paragraph. That's because linguists don't consider symbol-use of this sort as honest to goodness language. That in turn is because the symbols used are hardwired into those animals, whereas full blown language uses symbols that are totally arbitrary. And I add *"totally"* because Vervet symbols do seem *somewhat* arbitrary—at least when viewed at a species level. For instance, I could readily imagine call number one being used as a symbol for danger number three and vice versa, so these calls seem not as deeply embedded in nature as, for instance, smoke being a sign of fire.

But in principle at least, we humans can use any sound we want to stand in for any object we want it to stand in for. If I cared to, I could even invent a bunch of symbols in that manner for my own private use—as in fact the occasional patient of mine has done over the years (the better to talk to an exclusive audience of one, though eventually accepting me into this privileged circle). I myself have such a language, and I'm about to unprivatize a bit of it right now. "Ludny eckno rutni habbee" may be roughly translated into English as "The philosopher, Wittgenstein, has many admirers, but he has not always been right."

Though a person often enough uses language to clarify his own thoughts, we ordinarily *learn* language from others. And in a highly complex social species like *Homo Sapiens,* the complex exchange of information made possible by language has been the *most* crucial tool in our advancement.

Chimps, we should note, have some ability to learn and use arbitrary symbols, but that happens only slowly, to a small extent, and with a great deal of coaching. Two year old members of *Homo Sapiens,* by contrast, will pick up new words on a daily basis and will start putting them together into meaningful strings spontaneously—despite the fact that we haven't even bothered to provide them with a textbook in "Beginner's Grammar."

Nevertheless, the fact that Chimps do have rudimentary ability when it comes to the use of arbitrary symbols is enough to suggest the gradually building complexity in symbol-use among primates. And on that point, I'd like to put in another plug for my dog, Shep, who understands at least functionally a number of my words. When, for instance, I say we're going to "check the mail," he knows darn well *that* means we're going to head out to the road and check the postal box. And though he hasn't mastered English yet, he does seem to use certain arbitrary noises innovatively as symbols. For example, he barks in various ways when he's outside. But he uses one peculiar brief and muffled bark only at certain times, conveying something in the direction of "I wanna' come in now." Since the sound is distinctive, since it's *about* coming in the house, and since no other dog I've had has ever used this rather arbitrary sound, the instance leans toward real language.

And though dogs are less intelligent than chimps overall, they've evolved for millennia now in close relation with us humans, so it's not totally surprising that individual canines are capable of using the occasional idiosyncratic symbol to communicate with their human families. (These symbols are to be distinguished from species-wide symbol productions, as when dogs prance their front legs straight out and down as a way of saying "let's play".)

FROM PROTOLANGUAGE TO LANGUAGE

The use of arbitrary symbols as stand ins for objects and their motions is necessary for language production. And we can readily imagine scenarios where some forerunners of our grand and glorious species became skilled at using *already existing* behavioral-reinforcement mechanisms to associate a given sound with some external object. For instance, hominid _A_ emits the sound 'yrreb' and points in the direction from which she's just come. Another tribe member, hominid _B_, whom she's approached, starts to follow and lo-and-behold, a hill of blueberries comes into view. So hominid B's nervous system links the sound 'yrreb' with the positive reinforcement of berries. And from then on, any time his partner says 'yrreb' to him, he thinks 'berries'.

In part two of the above scenario, other linguistic pathfinders begin to extend this method of linking arbitrary sounds with specific reinforcing situations. For instance, the extremely primitive sound 'hubba hubba' became associated with a pretty female of their species walking by. And on and on until this group of hominids built up quite a repertoire of sounds linked with specific kinds of objects and happenings. Eventually, they developed what we late comers on the scene might refer to as an "extensive vocabulary."

Mid twentieth century behaviorists often touted this sort of procedure as the basis of language. That's what the great behaviorist, B.F. Skinner, taught, and he argued the position with political pundit (and sometime linguist) Noam Chomsky. Skinner lost the debate. But before explaining why, I'd like to reiterate a point I made earlier about the crucial importance of naturally selected behavioral 'programs'. In the hominid story just related, A not only makes a specific sound, she adds a visual symbol to the pot. As I put it, hominid A "points in the direction" of the berries. Now what she actually did at the time was to raise her arm till it paralleled the ground along a certain angle, and clasped her fingers together behind her overlapping thumb—all but her index finger, which remained unflexed. Why then did I assume in my story that B read this to mean "over here in the direction designated by my index finger." Or to restate the issue, how is it that hominid B would read the action as a *symbol*, in this case a symbol *about* direction?

The fact is, we take our amazing ability to read behavioral symbols so much for granted that we don't often reflect on the fact that doing so requires the presence of some pretty sophisticated behavioral programs. My sheep, for instance, don't seem able to read my pointing symbolism at all. So when I point in a certain direction, they never turn their heads *thataway* as a fellow human being would do almost automatically. Instead, they look more closely at me—at the source of the bodily movement.

And at this point, I have to confess with some embarrassment that Shep wasn't exactly a canine genius when it first came to pointing time. It took him a while to learn, although some dogs will pick it up first time out. The lesson? Seeing something as a symbol constitutes a highly useful skill, but it requires the right sort of 'program'. We members of *Homo Sapiens* have fancy version 10.5, while many animals have only version 1.0—and many other animals don't seem up to running this sort of program at all.

SKINNER VS. CHOMSKY

Back to the notion of language learned by behavioral conditioning. I said above that Skinner lost his argument to Chomsky. Behaviorists can still contend (rightly in my opinion) that reinforcement forms a useful mechanism when it comes to the development of basic vocabulary, but Chomsky won the scientific community to his general position by emphasizing one simple fact: small children readily learn how to use sentences that have *never* been reinforced a single time—for the reason that children create brand new sentences they've never heard before, and obviously you can't *reinforce* a behavior until it has actually occurred. Furthermore, children will start performing this incredible feat at a predictable developmental stage and without any formal teaching, indicating a built-in capability that unfolds readily at the proper time (just as there's a 'motor program' that underwrites baby's first steps during a predictable developmental period). Falling back on that old saw, "The devil is in the details," we should hasten to add that Chomsky has never been able to convince all scientists in the field that our built-in language capability involves his specific suggestion of a "universal grammar" program.

LANGUAGE—THE REAL McCOY

I said earlier that the use of arbitrary symbols is necessary for full blown language. But it's not sufficient. Symbols applied to objects in our environment may well have constituted a first crucial step—since there's no such thing as fossilized sounds, we will probably never know for sure—but there's a lot more involved in language, at least as linguists use the term. Namely, *syntax*.

If a language speaker says "I picked a bunch of berries up on blueberry hill," he's referring not only to physical objects (berries) but to their position (blueberry hill), to movements of the objects (they got "picked"), and who did it (the very special object called 'I'). Furthermore, all these symbolizing sounds have been woven together into a meaningfully connected stream. The use of words alone amounts to no more than a *protolanguage*. And chimpanzees, the species that currently gets second prize for adeptness-with-symbols, have one devil of a time moving on from the use of arbitrary symbols (words) to the much tougher problems involved in syntax.

Not to brag, but most members of our own grand and glorious species can follow even convoluted syntactical forms, and they can do so with one hand tied behind their backs. Take for example the song by Steve Goodman about men and women and sex. Here's how he put it: *"There are men who like woman who like men, there are woman who like woman now and then, there are men who like men and can't pretend, that they are men who like women who like men."* There's a whole bunch of syntax going on there, and you wouldn't have a clue as to what ole Steve was getting at if you couldn't follow syntactical rules. Yet I'll bet you had no difficulty at all.

OBJECTS, PERCEPTUAL AND CONCEPTUAL

So far, we have noted that our detection apparatus in effect forces us to experience discrete objects and their movements (even formless blobs will be seen as…uh…*blobs)*. Sophisticated mechanisms of this sort developed millions of years ago, so much so that *objects* became the perceptual coin-of-the-realm among all the higher animals, and it was upon this base that our extremely elegant symbolic apparatus became superimposed. Not surprising under these circumstances that just as our perceptual systems form objects, so does our conceptual system. This fact—usually taken totally for granted—shapes the way we think, and also limits the modes of thinking we can call upon. We will turn next, therefore, to an exploration of this crucial realm: conceptual objects and *their* movements.

6

CONCEPTUAL OBJECTS

Though we communicate in various ways—by expressive movements, for instance, or by drawing pictures—let's jump right to the honcho of our symbolic systems, our conceptual apparatus. When we think *conceptually*, we use words. You could demonstrate that fact to yourself right now by trying to think *without* words. I don't believe you'll have much luck, unless you substitute images and such. That's exactly what Einstein did at times in the process of fashioning his Special Theory of Relativity (e.g. he imagined himself riding a beam of light). Nevertheless, at some point he had to formulate his thinking in terms of words and mathematical symbols, manipulated in accordance with strict (logical) rules.

What about those occasions when we say something along the lines of "I know what I mean; it's just that I can't seem to explain it in words right now." I think our trouble on such occasions is that we *don't* really have a formed thought at all, but rather a hazy jumble—though it sometimes turns out that when we work on the hazy jumble, we're eventually rewarded by a clear thought that can provide us with innovative insight. But to have a clear thought *is* to be able to put it into words.

So intimate is this connection between our concepts and our words that we might be tempted to say: "Thoughts and words turn out to be the same things." But this effort at economy-of-explanation breaks down pretty fast, because I can have a concept—as a nonverbal stand in for the one presently in mind, I'm picturing the pond in my back meadow—and I can express this concept by either the word 'water' or the word 'aqua'. Same concept, different words. That convinces me that concepts are *not* the same as words. Yet, I don't believe I've ever really possessed concepts without the inexorably connected capability of expressing them in words.

How about those times when we have a concept but we can't think of the word to hang on it. Happened just the other day when I wanted to use 'tautology', but I couldn't for the life of me resurrect that word into consciousness.

Talking to my companion, I found myself stumbling around the point: "The word I'm looking for means that you're just referring to the same thing in different words…like a rowboat is a boat you row, or a square is an enclosed figure with four sides of equal length, where all the inside corners are blessed with right angles."

In a way then I *did* have the concept, just not the word for it. At least I had it clearly enough that I was able to come up with *a whole bunch of words* to illustrate the concept. Yet I knew those illustrations weren't exactly the same as the concept itself, because they were too concrete. They required my companion to generalize from these instances in order to understand the more abstract notion I was trying to get across.

To summarize then: There *is* something we call a concept, and it's not the same as a word, although we need words to convey our concepts. And not only to other people but also to ourselves. Words and concepts within our human information-processing equipment are as inseparable as the two sides of a coin—words forming the side that's face up, so to speak.

PERCEPTUAL OBJECTS vs. CONCEPTUAL OBJECTS

That being the case, we will concentrate now on the words we use to specify and convey our concepts. Words are individual things, usually in the form of sounds with a beginning and an end, or in the form of written symbols. In our Western languages, written words are composed of clustered letters (for example, any of the separated sequences of letters you see on the page before you now). They *are* objects, in the sense that they form discrete individuals you can see or hear. Yet words differ compellingly from ordinary objects in our surroundings. Physical objects are what they are. Period. Whereas a word, in so far as it *is* a word, designates something beyond itself.

THE FIRST WORDS

It's likely, as I mentioned earlier, that the first words used by our hominid predecessors symbolized objects in the environment. Fittingly then, each basic *conceptual object* would become <u>linked</u> with a particular type of *physical object* or event. But before we discuss the linkage in more detail, let me tell you a story:

This young girl says to her father: "Dad, you've got to do something about those brothers of mine. Their language is getting to be just awful." Well, naturally the father's upset when he hears about this, so he promises to straighten the boys out. Very next morning at breakfast, the older brother says to his sister: "Pass me the damn Cornflakes." The father immediately reaches across the table, and in his outrage, he slams the boy so hard that he goes sprawling across the room, landing on the floor against the opposite wall. Then the father turns to his younger son and says "Now, did you learn anything from that?" The little boy nods his head in fright and answers "I sure did; I'm never gonna' ask for the damn cornflakes again."

If a reader happens to thinks this story's worth repeating, let me mention that its lesson comes across more vividly if you replace "damn" with the F word. But in either case, the point is that though behavioral reinforcements *do* work, the particular element of behavior that gets associatively linked with the reinforcement isn't necessarily the one you expect.

So let's go back now to the story of the two hominids and ask how $\underline{B}$'s nervous system so unerringly linked the sound 'yrreb' with berries. Why did it not forge a link with the notion "Walk this way," or for that matter with the west wind that had just come up? Or with the early sunset that was starting to spread across the sky, or with the hill on which the berries were located, or with the bushes that were caressing the blueberries? And on and on to *any* of the linkages that $\underline{B}$'s nervous system might theoretically have forged with the sound.

That sort of concern can greatly trouble philosophers—given their tendency to focus on whatever is logically possible—but behavioral programs of social species provide constraints that usually limit such runaway linkages. Note, for instance, that the Cornflake incident's humor depends on the actual implausibility of the linkage reported. That's because we know young boys already possess a fund of knowledge cluing them in on the fact that Cornflakes are good (all those glowing attestations so magnanimously provided by the Kellogg company) and also that certain words are bad.

Back now to the Hominid Story: Based on the behavioral programs with which that early species was likely endowed, $\underline{B}$ would tend to associate the sound 'yrreb' with a mid size object that was strongly reinforced—in this case, the berries that tasted so yummy. But now let's change the scenario a bit. Suppose $\underline{A}$ and $\underline{B}$ were romantically attracted to each other, and the most potent reinforcement came from the bonding that took sudden root at this particular locale. $\underline{B}$'s nervous system might well then have linked the sound with the magical venue, *Blueberry Hill,* the place where the extreme positive reinforcement occurred. Which leads us to the issue of proper names.

PROPER NAMES vs. GENERAL CONCEPTS

Sometimes, a sound can be linked to a specific place or person, in which case we refer to it as a proper name. But over time, given a drift toward generality, words tend to categorize a *type* of thing (as in "Dear John, I think the world of you…"), rather than one specific object. Another instance: When we hear the old song, *"Blueberry Hill,"* we think of this phrase as naming a specific (albeit fictional) place that provides the song line's setting. But in another context, the phrase 'blueberry hill' can designate *any* hill that happens to be rich in blueberries. And to further illustrate how wide-ranging our use of symbols can become, *"Blueberry Hill"* might also refer by metaphorical extension to any equivalent place where a listener has had that sort of recurrently unique romantic experience. In fact, hit songs almost always touch people in this way, providing variants on a listener's own experiences—or on what they hope to experience, or on what they fear they might experience.

CONCEPTUAL OBJECTS *et al.*

We have seen that words/concepts are *symbols* standing in for other things. I have suggested that the first words might have been used to designate physical objects in the environment (e.g. 'berry'). We also have action words (e.g. 'pick') or state-affirming words (e.g. "The berries <u>are</u> ripe"). Not to mention words that *modify* the main words, as in "<u>delicious</u> berries" or "Blueberries will ripen <u>quickly</u> in this weather." And we have a whole bunch of other words to symbolize things like direction (e.g. 'to', 'from'), and so on. But all these accessory words revolve around the *objects* we perceive (and what happens to them), so we will spend most of our energy focusing on these object words.

THE MAGIC OF *VERBAL OBJECTS* (WORDS)

I can't resist another aside first, however, having to do with the *magic* of words. We're so used to using words that we don't often reflect on how amazing they really are. 'Blueberry' refers, remember, not to one specific berry I may be hold-ing in my hand. The word applies to *any* blueberry, *any*where, *any*time. Because of this far flinging property, Scholastics used to refer to such words as 'universals'; and knowing the importance of happiness in our lives, these philosophers even went to the trouble of using *universals* to create a joyful story. Since I love this kind of story myself, I'm now going to share their account with you. (I have to assume, by the way, that you have the same trepidation I have about death and the same enthusiasm I have about going on living; because if that's *not* the case, you'll find this story boring instead of happifying.)

At any rate, despite George Santayana's pessimistic conclusion that "the fact of our having been born does not auger well for our immortality," traditional philosophers were able to give us a more humanly satisfying account of the situation. Here's the way it went:

Material things last for only a limited time. That's because they're extended in space, and to be extended in space is to have parts. The trouble with having parts is that parts can be taken apart. In fact, that's the final common denominator in the destruction of all material things. An example: Mount Everest is being worn down slowly by the weather; Mount Helena got blown apart suddenly from within; but in both cases, their eventual demise involves the process of their parts being taken apart. So what about us? Aren't we physical objects extended in space, and don't we get destroyed at the end of our lives like all the other physical objects around us?

Here's where the Scholastics could smile reassuringly while they brought us up to speed on the happy implications of 'universals'. Granted, they would say, that all material things—our own bodies included—eventually get taken apart. But there's something extraordinarily different about universals, and we can see this fact clearly if we take the time to examine one. Let's take the concept 'size' to doubly emphasize the point. Though this concept is *about* size, it has no size. It applies to something as large as a mountain or as small as a mountain flower. Mount Helena has certain specific dimensions (or it used to before it got blown apart), and so does a mountain laurel growing on its new found soil. Yet the concept 'size' has no specific dimensions; or to state the fact even more emphatically, it has *no* dimensions at all. *None!* And since matter is, by its very definition, extended in space, the conceptual object 'size' (or any other universal) cannot be composed of matter.

Pursuing this line of thought, the Scholastics then asked how a material object, extended as it is in space, could produce something so superior, something so *totally different* from itself, namely, a *non* extended item like the concept 'size'? That would be to rustle up something a material object does not have to give—equivalent to a pauper donating a million dollars to his favorite charity! So there must be something else in us, something that's *not* material, if we're to account for the production of non material things like universals. To wit, there has to be a *non* material principle within us to produce these *non* material products. The Scholastics, as it turns out, had a synonym for non material. They called it *spiritual,* and they concluded that we'd been gifted with a *spiritual princi-ple* (a spiritual soul) in addition to our material body.

Now we're getting to the best part: Since this spiritual soul is not extended in space like material objects, it has no parts. Recall now that the way physical objects get destroyed is by being taken apart. But with no parts, there's no possi-bility of this spiritual soul being taken apart. Hallelujah, we (at least *we* in the form of our spiritual souls) are never going to get destroyed. Only the dross of our physical bodies can go the way of all flesh.

Unfortunately for Santayana, however, he was unable to take this sort of account too seriously. That may be because, although a philosopher by trade, he probably still realized that all nervous systems seem to be in the business of gener-alizing from individual instances.

THE *GENERALIZING* TENDENCIES OF NERVOUS SYSTEMS

At *fin de siécle*—last time around—Ivan Pavlov was busy at work investigating the nervous system's ability to link bodily responses to external events. During the process, he showed in effect that the nervous systems of dogs clearly *generalize* from particular things detected in their surroundings, even if their systems don't hammer these representations into the verbal generalizations so congenial to our own species.

To illustrate Pavlov's findings, let's take a moment to outline one of these clas-sic "conditioning" experiments:

1.   Present meat to a dog > he will salivate (an innate reflex).

2.   Ring a bell > the dog will not salivate (innate reflex lacking).

3.   Establish a pattern of bell ringing followed by the presentation of meat > the dog will, as expected, continue to salivate.

4.   Now ring the bell, but don't present the meat > the dog will salivate, and do so as soon as he hears the bell.

Pavlov's explanatory story: The sound of a ringing bell has been *linked* in the dog's nervous system with the meat that evoked an innate reflex. That's why the dog displays the same kind of response to this newly <u>conditioned</u> stimulus.

Note now that the dog does not just respond to a specific piece of meat. His system automatically generalizes the response to *any* piece of meat (or even to meat powder). This generalizing function also occurs in respect to the conditioned stimulus, the ringing bell. Ringing *any* bell with the same sound will work. Shift to bells of different pitch and timber, and the conditioned response will generalize—up to a certain point, when it will start tapering and eventually disappear. The lesson: long before words came along, animal nervous systems were *generalizing* to beat the band, and were using symbols (e.g. representing the ringing bell as *about* something beyond itself) long before our 'symbol' for symbol came on the scene.

THE WONDERS, AND THE LIMITATIONS, OF WORDS

Still, even though our conceptual objects cannot act as unique guarantors of our immortality, they're pretty wondrous tools nonetheless. Like our perceptual objects, our words provide highly useful engineering shortcuts. However, as with all engineering shortcuts, there's a price to be paid. In the next chapter, we will highlight some of the advantages associated with our obligate use of conceptual objects. Later on, we'll get to some of the disadvantages.

7

THE WONDER OF WORDS

Even if all-that-is consisted of one overarching field of force varying extraordinarily in its intensity, such a physicist's view of 'what is' would have little immediate relevance for the ordinary world of our human experience where mid size objects and their movements reign supreme. We saw, for instance, that if our receptor mechanisms bundled extreme intensities of such field-forces into *perceptual objects,* the strategy would provide a highly economical way for the computationally limited human nervous system to represent and process its information. And the same efficiency applies when the resulting objects of our perception resonate with *conceptual objects,* as these items come into play at still higher levels of information processing.

To illustrate: I see a particular blueberry I'm about to pick, and the word 'blueberry' may well jump to mind. But this concept refers not just to the object I'm placing in my hand; it refers to all the blueberries that ever were or that ever will be. Verbal *generalizing* ability of this sort forms a spectacular achievement, because we can use such generalizations to understand the past and to predict the future. Even in this minor case, for example, I can move the concept 'blueberry' around in order to appreciate why people have gone blueberry picking for millennia, why raising blueberries has become big business in Maine, and why blueberry futures are not a bad investment. However, if 'blueberry' had referred only to the specific thing I just picked and no more, the sound would have slipped away silently, devoid of further application.

This blueberry illustration brings to mind, by the way, a true story that involved me' sainted muther when she was about five years old. At the time, all the children in her neighborhood had been warned by their parents not to go blueberry picking, because there'd been trouble with a pervert roaming the woods outside of town. You probably know how teenagers are, however, and a bunch of the older girls decided they weren't going to miss the height of blueberry season just because their stupid parents were being overly cautious. My mother begged her oldest sister to take her along—which my aunt Ann did, but only after extracting a solemn promise that my mother would never squeal. All went well during the excursion, and when they got back to the corner of their street, my mother saw my grandmother on the front porch. Running happily into her arms, she said: "Mommy, mommy, we weren't blueberry picking." With a family history of such wretched incompetence at lying, you can see why I try to avoid this vice whenever possible. Still, there's a difference between lying and telling a simple story. So let me now provide an account of the origin of language that starts with an optimistic 'thumbs up'.

HOW TO THROW WORDS TOGETHER FOR GOOD EFFECT

We've noted that language does not consist merely of individual words. We combine words in ways that greatly extend their usefulness; but in order to have our words interact helpfully, we need rules for how one word will act on or be acted on by another one. The rules involved have to be derived in some fashion, and the 'opposable thumb' story I'm about to tell provides a homely account.

When it came to the evolution of hands, the formation of a thumb that could be placed opposite the fingers provided a big step forward. You can demonstrate this fact to yourself right now by manipulating the objects around you while *not* opposing your thumb to your fingers (I think you'll find your efforts rather klutzy). So useful in fact was this transposition that some theorists focused on the occurrence as *the* pivotal factor in expansion and enhancement of hominid brains. That's because fancy hands with opposable thumbs wouldn't have been of much use unless accompanied by the greatly expanded computational equipment needed to underwrite fine digital movements...

Let me interrupt my story for a moment to emphasize that this crucial qualification was reinforced for me recently after a patient of mine suffered a small stroke. When she got up in the morning, she knew vaguely that "something just wasn't right." But the first specific deficit she recognized was an inability to button her blouse, a complex act requiring a *tour de force* in fine neural control (albeit we're so used to this wondrous ability of our normally functioning brains that we tend to take it totally for granted).

Back to the story. If our more immediate predecessors were to deftly hold and accurately release a rock or stick for hunting purposes, their brains needed to develop greatly enhanced computational skills. And not only in regard to digital dexterity. Hand movements needed to be coordinated with arm, shoulder, and body motions if a missile was to be effectively projected. And there was more. The newly programmed hands encouraged even further brain expansion, allowing *inner representations* now of innovative items that might be constructed with this fancy new apparatus. By the time David and Goliath came along, for instance, Goliath's fearsome spear was more than a casual stick he'd picked up along the way to battle, and David wasn't just flinging a stone. In short, the 'opposable thumb' tale provided a fulcrum for explaining the wondrously developed brain that was ultimately achieved by our own grand and glorious species.

Only problem with the story, other than its simplicity, was that Nature has no mechanism for establishing devices that *may* become useful at some future date. If an early predecessor of ours had humanoid hands with opposable thumbs but lacked the computational apparatus to underwrite digital dexterity, then the hands would be just hanging around, so to speak, amounting to no more than biologically expensive boondoggles. To wit, wasting energy on constructing and maintaining fancy hands of limited utility would make the organism *less* likely to survive. Natural Selection comes into play only when a random feature leads to adaptive advantage for an organism during its own lifetime. That's because the chance feature has to provide the organism with a leg up in regard to its own personal survival in order to account for progressive reproduction of the new feature.

So we need to make an alteration in our original story. Forelimb appendages and brains had to evolve gradually together, racheting each other along. As it eventually turned out, human-like hands became for our immediate predecessors the primary *mechanical interface* with the material world. Arriving later on the scene, language became for our own special species the primary *cognitive interface* with our social world. And language in many respects followed the type of control programs already laid down by the preexisting mechanical interface. Or as F.R. Wilson put it, *the human brain seems predisposed to generate language rules that treat nouns as if they were stones and verbs as if they were levers or pulleys.*

To illustrate: "We pick the berries and put them in buckets. We carry the buckets home and place the berries in bowls. Then we add fresh cream, and we eat them." And by the way, if I'd paid enough attention here to rules of English by throwing the pronoun *'them'* closest to the object it was standing in for, I wouldn't have confused those who don't yet know that you eat berries, but you don't eat bowls (fortunately, context so often bails us out).

OBJECTS CAN BE CATEGORIZED IN MANY WAYS

Everyday language makes heavy use of *conceptual objects* that reflect features of the *perceptual objects* around us, along with the manipulation of these objects, as in "We pick the berries". A fancier feature of conceptual objects, however, is their neat ability to reflect, not only features of external objects, but also features of our other conceptual objects—and they can do so in an indefinite number of directions. For instance, we can start with 'blueberries', 'strawberries', 'raspberries', and we can attend to only some of the features they have in common. Then one word fits all, namely 'berries'. Now we focus on features berries have in common with apples, peaches, pears, and plums, and we make up a word like 'fruit' that encompasses all these conceptual objects. Then we focus on features that the conceptual object 'fruit' has in common with string beans, roast beef, and scrod. We can then apply the word, 'food', and on and on.

I know, by the way, that talking about language stuff tends to become pretty tedious, pretty fast—except in Boston, where people can get carried away with the niceties thereof. Here's a case in point: Fellow comes to town and tells the Boston cabby that he's been looking forward to arriving at this famous port city because he's been wanting to get scrod…and can the cabby bring him to a suitable establishment. "Sure," says the driver, "but you know, I never heard anyone use the future pluperfect like that before."

Back to the fact that we can form conceptual objects that reflect features of other conceptual objects in an indefinite number of directions: We could also bundle blueberries, the Old State House, Cape Cod, and the Berkshire Hills together—under a concept like 'items found in Massachusetts'. And though we can form an endless number of categories, the actual ones we use connect with our human interests. We haven't bothered to provide a word referring exclusively to sneakers-cameras-clouds-and-thermometers as members of the same special category. But we do have such a label for tables-chairs-couches-chests-and-beds, namely 'furniture'. The latter category suits our human interests, while the former grouping would have no practical use.

Of course, "human interests" vary from society to society; hence, also the words that get generated. To illustrate, in cultures where exact family relationships are crucial, there will be more words to separate off the different relationships. And even though such exactitude is not overwhelmingly important in our own culture, the poverty of available categories can still lead to temporary ambiguity. A number of times over the years, for instance, I've been confused about which specific person my patients were referring to when they mentioned, say, their "brother-in-law." Especially when such patients came from large families, I've not been sure at first whether they were referring to their sister's husband, their spouse's brother, or their spouse's sister's husband. At times, patients have even extended that phrase to their sister's husband's brother, their spouse's sister's husband's brother, their spouse's brother's sister's husband…and I think I'll stop there. In principle, we could of course invent a different name to designate each of these categories explicitly, but in our culture, it wouldn't be worth the work. (Guess I could've used the above case though as an example of where there's a whole bunch of syntax going on.)

TWENTY QUESTIONS

The old parlor game, Twenty Questions, illustrates how conceptual objects help us organize our information and focus our attention. We start with three categories—animal, vegetable, and mineral—each broad class formed from features found in more restricted conceptual objects (ultimately deriving from specific types of perceptual objects). Take 'animal' for example. We begin with perceptual objects having features that allow us to segregate dogs from cats from buffaloes from armadillos.[2] Then we select features that are common to all, and label the very encompassing cluster 'animal'.

One of the participants in this parlor game will privately choose an individual object or type of object, and then will answer whether the object is animal, vegetable, or mineral. Just this three-part division narrows the field remarkably for the other contestants. From then on, the secret-holder will limit her response to questions with a "yes," or "no," (e.g. "Does the object fly under its own power?"), and remarkably, the other contestants are quite often able to zero in on the exact object within the twenty questions permitted.

But what if you were to choose 'quark' as the secret object? How would you then answer the question: "Animal, vegetable, or mineral?" The game breaks down once objects are chosen that do not conform to the three broad categories enfolding our *ordinary world*. Nevertheless, the point remains: our ability to use general words has enabled our species to organize things in ways that have gifted us with enormous power.

BEST OF ALL

Best of all, brain encodings involved in the formation of conceptual objects get to hang around long after the perceptual objects themselves have gone away. I've been talking about blueberries, for instance, but I won't actually see one on my land for another few months. Yet I can conjure up 'blueberry' now without difficulty, and combine the concept with others like 'pancakes' or 'muffins'. I can even combine it fancifully with a concept like, say, 'castle', and picture a vast edifice of blueberries rising in the distance—though I can also lay the concept of 'tensile strength' right next to 'blueberry castle', look at the two notions together, and then judge that they don't fit awfully well.

This ability to move our concepts around and put them next to one another in jigsaw fashion can be lifesaving. Reminds me of my college roommate's account of his first car, a hot rod that he bought as a senior in highschool. He'd worked hard, saved his money, and found a good buy (no Hollywood mufflers, but a small hole in the conventional one created a similar effect). He was so excited that the first time he backed from his driveway, he revved the engine and fairly rocketed out to the street. Only then did he break into a cold sweat. That's because the thought of another car zipping along the road pushed itself right up against the mental image of what he'd just done. From then on, Bill restrained his enthusiasm a bit—all of which illustrates the dictum that it makes sense to let our imaginary productions die in our stead.

ALL-PURPOSE PRINCIPLES

Our use of language, in which we're able to select features from conceptual objects, then select features from *these* conceptual objects, and repeat this process as many times as we want (e.g. blueberry > berry > fruit > food > thing), allows us to arrive at extremely general notions, including principles that we can apply to everything that is. Such all-purpose principles extend our human manipulative abilities remarkably, so let's take a look next at the usefulness and limitations of our so-called "First Principles."

8

FIRST PRINCIPLES

I don't know if you've had any direct dealings with angels, but some of my friends tell me they have. Theologians assure us that these creations of the Almighty far surpass mere humans like ourselves, because angels are pure spirits, unhampered by the dross of our material bodies. Still, concerned theologians had to figure out how angels got to be so smart, given the fact that they had no sense organs to gather information. The experts finally concluded that God must've provided angels with a whole bunch of 'infused' knowledge.

In stark contrast, we have to get our knowledge the old fashioned way. As the saying goes, "We work for it." The empiricist, John Locke, went so far as to judge that each of our minds enters the world as a totally blank slate *(tabula rasa)*. Then our senses get to work, providing us with every stitch of information we eventually grab hold of. Some Scholastics thought differently, maintaining that we don't have to work for *every* scrap of our knowledge. They believed that God infuses some crucial innate stuff. So we don't have to be taught, for example, the *Principle of Contradiction*, or the *Principle of Causality*, because such First Principles have been built into the human soul. And as far as the reliability of these Principles is concerned, an all-good God would never have stashed the wrong gear in our minds.

That conviction made Scholastics a bit less concerned when David Hume came along to underline the fact that nowhere do our senses provide us with, say, the concept of *causality*. All we see is that one thing happens, and then another thing happens right afterwards. A cue ball hits the stack, for instance, and next thing you know, pool balls are scattering all over the place. But there's nothing visible to the senses saying that the cue ball *caused* the other balls to scatter. So how do we really *know* that our Principle of Causality always accurately reflects the real world outside our own thoughts?

That was an important question, because if you were a person who didn't take an all good God for granted—laboring instead under a felt need need to prove that 'He' existed—this meant big problems. Why? Because the reasoning involved in the proof of God's existence then became rather circular. One would be applying, say, *The Principle of Causality* in order to prove that God exists, but at the same time assuming God's existence in order to justify the reliability of *The Principle of Causality*. Okay for the Scholastics to do, since they accepted God's existence anyway as a matter of sheer faith, but not for the many doubting Thomases looking for rational proof.

KANT'S PROPOSAL

A while after Hume revealed his skeptical ponderings, Immanuel Kant came along to provide an explanatory story as to *why* our minds inevitably think in terms of principles like causality. The human mind, he said in effect, is programmed in such a way that we cannot help but process information in the manner we do. Kant's notion, it must be said, provided scant consolation for Guardians of the conviction that we can indeed know what really is, and a moment's reflection will help us see why. For if a principle like *"Every thing must have a cause"* comes from the human mind rather than from the world around us, might it not be the case that concepts like 'causality' tell us a whole bunch about our own minds, but not a heck of a lot about the real world?

Kant played out his entire life, of course, in the dark ages B.C. (i.e. Before Computers), and therefore did not actually think in terms of computational programs. Instead, he spoke of innate categories of the mind. But nowadays, theorists armed with knowledge of modern cognitive science do not think of 'mind' as some sort of separate entity, but rather as a shorthand term to enclose all the cognitive activities produced by the brain. Modern theorists also recognize that brains have numbers of built-in programs useful for adapting to a species' *ordinary world*. So modern theorists think less in mentalistic terms like 'Kantian Categories' and more in functional terms like brain circuits that come with a generous portion of prewiring.

Further, instead of postulating a Divine Person as the *immediate* Guarantor of a particular program's inherent correctness, most modern theorists apply the concept of Natural Selection: If an inherited program does not prove useful—if the program's output leads to maladaptive behavior—then the organism, along with its programs, will not likely survive, let alone reproduce. Note carefully, however, the small print at the bottom of this alternate guarantee: *Natural Selection assumes no responsibility for any claim that a particular cognitive program will provide "the truth, the whole truth, and nothing but the truth."* The guarantee states only that if an organism applies the program within the limits of its *ordinary world*, the results are highly likely to aid in successful adaptation within that world. (This caveat will be important for us to keep in mind as we go along).

CAUSALITY

So what about, say, the Principle of Causality? Does that amount to a program hardwired into our brain/minds? In attempting to answer this question, we need to avoid what William James referred to years ago as the "psychologist's error"—confusing our way of *thinking* about things with the actual mechanisms involved. If, for instance, we stated the Principle of Causality in one or another abstract form, say, "All things must have a cause," then there may well be people who go through their entire life without *ever* using the "Principle of Causality."

In our daily lives we do, however, use many action words. We push, shove, thrust, throw, heave, press, squeeze, crush, chew, pull, tug, drag, carry, haul, *et cetera*. And since we view all of these activities as bringing about change, we could say that each one represents a type of *causal* activity. But we usually don't conceptualize at such an abstract level. In fact, we'd think it rather odd if a person said "I *cause* the berries to leave their branches" rather than "I *pick* the berries." Why? Not because we'd consider it technically incorrect, but simply because folks don't usually express themselves at an unnecessarily abstract level. We expect, for instance, that the fellow next to us may mention that he *drove* or *took the bus* to work, but not that he *transported* himself to work. In fact, "transport" leaves out too many details. Did he walk, run, ride his horse, take a bus, or use a hot air balloon?

'PUSH' is to 'CAUSE' as 'BERRY' is to 'FOOD'

Recall that one of the neat things about manipulating concepts is that we can center on a property common to a whole bunch of them and make up a new category composed of concepts that have the chosen property. That's how we got from 'blueberry' to 'berry' to 'fruit' to 'food'. Similarly, "push, shove, thrust, throw, heave, press, squeeze, crush, chew, pull, tug, drag, carry, haul," share a common property. In all cases, there's an implication that *one object is bringing about change in another object*. All the action words I've just mentioned share this property, and we can label the property 'causality'. But ordinarily our minds work at a more concrete level—very concrete early on, for what's available to infants is not an abstract concept, but rather a working sense that one thing can bring about change in another thing.

THE FUNDAMENTALS OF CAUSALITY

To make this point clear, let's take a look at the following stripped-down experiment concerning human perception of causality. Each participant in the test watches a TV display of three different sequences: **1.** A red ball enters from one side of the screen and moves to the center where it touches a green ball, which immediately moves the rest of the way across, exiting on the other side. **2.** Similar sequence, except that when the red ball touches the green ball, there's a significant time delay before the green ball starts its journey. **3.** Similar sequence, except that the red ball stops before touching the green ball, though the green ball starts its journey immediately after the red one has stopped.

Which of these three arrangements would you rate as a *causal* sequence? Participants in the actual experiment rated the first sequence as causal—the red ball *hitting* the green ball and causing it to move. But when the red ball stopped at the green ball, and the green ball did not move for a while (sequence 2), or when the red ball never got as far as the green ball, even though the green ball started off immediately afterwards (sequence 3), participants tended to rate the latter two occurrences as *non* causal. Now here is the really interesting point: when the same test was given to infants about six months old, they voted the same way, suggesting that we do indeed have a built-in sense that one thing can bring about change in another thing.[1]

"Wait a minute," I can hear someone asking, "last time I checked, infants of that age weren't up to talking, never mind following directions, so how'd they manage to take such a test?" Turns out that while folks in business have been wracking their brains to figure out successful financial plans, developmental psychologists have been wracking *their* brains trying to figure out successful testing procedures for little ones. And what scientists eventually came up with was the ingenious application of a phenomenon that occurs also in us adults: When something pops into view that's novel, unexpected, and interesting, we tend to look at it for a longer time. Newly wrecked vehicles on the roadside, for example, lead predictably to an ongoing traffic creep, and we say that the resulting tie-up is due to the "curiosity factor."

In like manner, show infants something new on a TV monitor, and they'll look with interest for a period of time. Repeat the same picture until it becomes tedious, and they won't bother looking very long. They have become habituated to the scene. Now show something new, and they will dishabituate; that is, once again they will look for a longer time. This 'Habituation Paradigm' is what developmental scientists have used to test infants with sequences such as the three described above.

Let's view a brief sample from tests performed on infants just a bit older than six months. Show one of the two 'non causal' sequences noted above, and repeat the display until infants show only a baseline interest, as judged by how long they continue to look at each recurrent display (less than six seconds on average). Then show one of the other two sequences. If it's the non causal one, infants will look a couple of seconds longer at the new variation. But if it's the *causal* sequence, infants will look twice as long. Why? Because infants see this event as something qualitatively different from just another similar non causal sequence.

We should perhaps emphasize: Even this basic 'causal program' requires a number of months to get up and running in infants, because the developing nervous system has first to consolidate its characteristic way of cordoning off stimuli into discrete *objects* that move *in* space, *over* time. Why this step is necessary becomes clear in the context of that old sports dictum: "You can't hit what you can't see." Only in the present case: you can't conclude anything about objects in motion until you can see *objects*, and perceive their *motion*.

Over the several months following the initial assessment of causality, infants begin to respond to some of the finer points. For instance, they start to readily distinguish the active object that causes change from the passive object that gets changed. And they distinguish different varieties of physical causality like pushing and pulling. And as toddlers start to use action words, they hammer home such distinctions in our characteristic human (i.e. verbal) way.

DO OTHER ANIMALS KNOW ABOUT CAUSALITY?

If 'knowing' about causality means having the verbal concept of causality, then obviously the non verbal animals don't know a darn thing about causality. But functionally, Shep sure knows about causal transactions. If he's confronting a sheep whom he wants to move to the (sheep's) right, Shep will take a step to his own right and slightly forward, eyeing the animal intensely. The sheep will move to its right in response (heading in the opposite direction). Then Shep will take a measured step toward the sheep, who will almost always continue then along the path it's started on to its right. And I say "measured step" because Shep does not want to panic the sheep into erratic flight. Instead, he follows from behind at just the right pace.

A traditional behaviorist might well say that such behavior tells nothing about Shep's knowledge of causality, and that learning-by-association can satisfactorily explain the sequence. For instance, Shep perceives that his motion in one direction is regularly followed by the sheep's motion in the other direction. And if he moves too quickly, his behavior will be negatively reinforced by the sheep scattering erratically. There's a tad of unintended irony in that sort of analysis though, because it brings us back almost full circle to David Hume's point that all we humans ever see is the same sort of sequence: first this happens, then predictably that, so we end up referring conceptually to the sequence as 'cause' and 'effect'. Perhaps we should say that we *know* about causality, but Shep has *'know how'* about causality. As do other non verbal animals, like the sheep who know how to avoid Shep's too close proximity—and the infants who are responding to a causal sequence many months before the verbally most precocious of them will have a word for it.

AWARENESS OF OURSELVES AS A CAUSAL FORCE

When Shep is exhibiting his *know how* about causation, he's moving his body, same as we so often do when we have the personal experience of using ourselves as causal agents. If we hammer a nail, for instance, we experience a sense-of-effort as we pummel it home. And we often will use a word like 'force' to account for such effortful changes that we bring about. But if someone were to ask us for a formal definition of 'force', we'd have big trouble providing one that wasn't circular. Look it up in your dictionary right now, and you'll find the usual strategy: First, the dictionary will provide synonyms like "exerting physical *power*," and then provide examples like "She used all her force to open the window." If you continue the dictionary game by looking up "power," you'll find some more synonyms like "strength, might, *force*." (And that's before the problem arises of using the word in less physical and more metaphorical senses like "moral force.")

My technical *Encyclopedia Of Physics* has an interesting way of dealing with the problem.[2] The word 'force' is, as one would expect, used liberally throughout the text. But when I went to look up the concept 'force' itself, I found an empty sandwich: entries for *"Fluid* Physics" on the one side and *"Fourier* Transforms" on the other side...and the notion of *"Force"* nowhere to be found between the two.

Things don't seem to have changed much since David Hume wrote that 'force' was one of the most mysterious concepts we have. It's too elemental, in fact, to enable formal definition in terms of more basic components. And yet we do have that immediate experience of effort as we close the sticking window or hammer the nail home. Then we take the same *subjective sense of force* and project it onto the inanimate things around us, even to agents when they're not visible. Here's an example poetically provided by Christina Rossetti: *"Who has seen the wind, neither I nor you, but when the leaves are trembling, the wind is passing through."* (That line pops to mind every time I see the leaves on my quaking Aspens begin to dance and twirl.)

So the notion of 'causal force' comes additionally from our own immediate experience, although that happens some time after infants already show a 'know how' concerning causality. (At that early age, they don't yet have a totally firm notion of themselves as fully separate objects/agents—but two year olds have the notion down pat, albeit sometimes to the consternation of their parents.)

DO 'FIRST PRINCIPLES' LIKE CAUSALITY *ALWAYS* APPLY?

Within our own *ordinary world* of midsize objects and their movements, the principle of causality seems to work extremely well. We will see in the coming chapter, however, that problems arise when we move beyond the limits of our ordinary world. And it's the same with the other 'First Principles'.

9

A SECOND LOOK AT FIRST PRINCIPLES

I related the 'Opposable Thumb' story in Chapter 7 to provide an easy-minded rationale for the occurrence of highly refined *patterning and sequencing* abilities in hominid brains—abilities that set the stage for progression to the intricate programs of this nature required for language. We could rerun that same story now, however, in order to make an additional point, this time about the way we tend to think causally: Our systems appear to focus whenever possible on *one salient attribute* from the vast and ever intertwining network of causes/effects (as in the above story's focus on one item), treating this anointed feature as THE cause of whatever we happen to be attending to.

Another illustration: When Presidents have the good fortune of holding office during prosperous times, their partisans will note that "under our President's superb leadership, our great Country is prospering as never before." Not so good for Presidents during a recession, however, because "under this President's failed leadership, our Country has taken a nose dive. Elect me, and I will bring about full restoration of our Land's inherent greatness." A President's financial policies do of course make *some* impact on the overall economy, but the present point is that we have a marked tendency to focus on one conspicuous facet of a complex situation as if it were *the* cause of whatever has occurred.

Strange perhaps that this oversimplifying tendency of ours can succeed at all, but it does have the virtue of helping us to avoid swamping our limited information-processing apparatus. Furthermore, the fact is that this engineering shortcut *does* work—at least, it works sufficiently that we often end up with highly useful approximations. So if Larry Bird scored over 40 points during an old Celtics game—including a three point conversion to put his team on top in the last desperate second—we can excuse sports writers for telling us that Larry won the game. Fortunate also, however, that the human brain can monitor this simplifying 'default' program, enabling us to become sufficiently aware of the shortcut involved to make adjustments when needed. We tend to do this especially when no causal aspect of an event stands out in obvious strength from the others. On such occasions, we often end up thinking in terms of probabilities.

PROBABILITY

The referee flips a coin to decide which football team will have an opportunity to receive the ball first. We know the odds of the coin landing either heads or tails are 50-50, and we assume that the ref doesn't know any more than we do what the outcome will be. That's because he doesn't know exactly how much force he's put into the flip. Therefore, he doesn't know exactly how high the coin will rise—hence, how extended a time the coin will have to complete its turns before hitting the ground. He also doesn't know precisely where the impetus of his thumb has struck the coin, so he doesn't know exactly how rapidly the coin will rotate. Further, he can't judge how much the swirling wind will change these dynamics before the coin hits the ground, or how the turf will effect the coin's roll before it comes to rest.

When faced with multiple causal factors, not fully specified as to strength, and interacting in ways that are not entirely predictable (*e.g.* the wind may swirl just as the coin turns with its broad face outward like a sail), we say the result is a matter of 'chance'. It's not that we've stopped thinking in terms of ordinary causality, simply that we don't have enough information to predict the exact result. We know, however, that there are only two possible outcomes of equal opportunity, so we can at least know the probability.

In our *ordinary world* of mid size objects and their movements, we use this sort of statistical shortcut often, and to good effect. Still, we take it for granted, at least implicitly, that we *could* know the exact outcome if we had enough causal information. To illustrate, think for a moment of the following coin-flipping apparatus: Our machine will be able to strike a coin with known force at an exact spot each time. The room in which the test is to be conducted will be engineered to have no appreciable air currents, and the flooring's firmness and elasticity will be made uniform. By accurately controlling these factors, we could arrange to have the coin come up heads or tails as we choose. Stated another way, we fully expect that <u>an identical constellation of causal factors will produce the exact same effect</u>. Since the cause will fully determine the effect, theorists refer to this type of inevitability as *determinism*.

QUANTUM PROBABILITY

Now let's move beyond our *ordinary world* of midsize objects and their movements to, say, the quantum world. The above illustration might seem like a good introduction to this extra ordinary world, given the fact that Quantum Mechanics deals ONLY with probabilities. But strange things happen when we pass through Alice's looking glass, so to speak, to enter the "weird" world of quantum phenomena—and I place *weird* in quotes because particle physicists seem so fond of using the word when they comment on their findings. "Weird" refers to how the quantum world appears to *us* humans when we try to fit the results into our ordinary framework.

Particle physicists describe the situation by saying that quantum reality is *"inherently* probabilistic," a simple phrase, but one with strange (to us) implications. We could use the referee's coin toss again to illustrate the profound difference. Recall in such a case that we use probability to get a handle on likely outcomes when we don't know enough of the causal details—but always with the sense that we could *in principle* know the exact outcome if we knew enough of the pertinent variables. Since we don't, we refer to them as *hidden variables*. But by controlling these factors as we did in the test apparatus, the variables become unhidden, and we can then calculate the exact result. Restating the point for emphasis, what we expect is that *"an identical constellation of causal factors acting under the same conditions will produce the exact same effect."*

But *weirdly,* that's not the case in the quantum world. Set up an experiment as precisely as one might want, then run it over again, and the results will likely be different—whether one is measuring, say, the time of the next release of an alpha particle by radium, or the half-life of a free neutron. Repeat the chosen experiment accurately for a hundred times, however, and the results will show a totally predictable probability curve, one that agrees precisely with quantum theory. And because no "hidden variables" can be detected to explain the statistical scatter of the results, particle physicists say that the outcome is *inherently* probabilistic.

Yet how can this be? For as we have noted, in our ordinary world, operation of an exact constellation of causes will always produce the same effect. But bewilderingly, in the quantum world, application of our Principle of Causality simply breaks down. Not that there are no antecedents to an event, but identical as the experimental setup may be, the antecedents do not work as our causal principle would require by producing a predictable effect in each successfully completed experimental trial. The amazing agreement between experiment and quantum theory occurs only at a statistical level, where the results conform precisely to the probability curve mathematically predicted by the theory.

That the quantum world does not follow the Principle of Causality as it's found in our *ordinary world* seems almost incomprehensible to us. And indeed such an abnegation seemed so repugnant to the great Einstein that it drew from him the famous complaint that "God does not play dice." In fact, forever after the formulation of quantum theory in the late 1920s, he (Einstein, not God) argued that somehow there must be "hidden variables" underlying the experiments. But not only were none ever found, subsequent theoretical attempts to interpret data in accordance with the expectations from our *ordinary world* have regularly failed—except perhaps for a rather forced 'Pilot wave' theory that ironically required information to travel faster than the limiting speed of light that had been the guiding principle in Einstein's own ground breaking theories of relativity.

PRINCIPLE OF THE EXCLUDED MIDDLE

Let's look now at another example of our First Principles, one that became highlighted in a lamentable Country Club event of recent vintage. Seems there was this highly attractive matron whose hardworking husband was away on a business trip at a time when the Club was throwing its summer bash. Naturally, her friends didn't want her to languish at home, so another couple drove her to the affair where she sat with her friends during the festivities. She was not up to staying to the wee hours of the morning though, so one of the husbands gallantly offered to drive her home after the banquet. As they were riding along, he innocently posed a theoretical question: Would she ever be unfaithful to her husband…one time only…for ten million dollars? She thought about it for a while, and ended up opining: "Well, John works so hard, and it would make our lives so much easier…and he'd never have to know…so maybe I would." A bit later, her friend swung off the main road, pulled over, and started to make advances. Naturally, she took umbrage. Pushing him away, she asked angrily: "What do you take me for?" To which he answered: "We've already decided *that*. Now all we're doing is haggling over the price."

Her gentleman friend had, of course, simply applied our logical *Principle of the Excluded Middle*. Either A, or not A. For as the Scholastic philosophers used to put it, *"tertium non datur"* (no third possibility is available). Granted, applying this principle can become a bit tricky at times, because you do have to be clear on your definitions. And those of us rushing to defend this attractive woman's honor would be quick to point out that "One swallow does not make a spring"—so we'd argue that her soon-to-be ex friend had placed her in a category to which she did not legitimately belong. Nevertheless, the Principle itself seems sacrosanct in our ordinary world. For instance, either the Red Sox will win the World Series this year, or they will not. Either Elvis Presley is still alive, or he is not. Either dinosaurs roamed our planet in the good ole days, or they did not. And so on.

PARTICLES

With this principle in mind, let's look now at *particles,* a subject that we glanced at briefly in Chapter 4. My American Heritage Dictionary says that a particle is a "very small bit," and gives as an example "a particle of dust"—which happened to be the same instance I was thinking of using by way of illustration. Such particles are sufficiently insubstantial that we ordinarily don't see them, at least not until a huge bunch of them accumulate on an undusted surface, or curl up into dust bunnies on the floor. But on those occasions when just a shaft of sun enters a room, we can see myriads of these tiny particles floating indolently through the beam of light. Each one has a definite size, diminutive as that may be. And moment to moment, a given dust particle occupies a specific location in space. So is it also with bread crumbs, bath powder, or any other particles in our ordinary world.

Particle physicists have used the same word to characterize their own specialty, a moniker that seems highly appropriate on first viewing, given the fact that these physicists are dealing with the teensey-weensiest of particles, those that are smaller even than the once-thought-indivisible atom. Let's use the electron for illustration, especially since most of us can visualize the atomic logo picturing electrons spinning around an atomic nucleus. This is the planetary model of the atom, so-called because electrons are represented as discrete little 'planets' moving around the nucleus (the sun). This *representation* might still be useful perhaps in an introductory chemistry course, though for a number of reasons that would take us too far afield, physicists have long since abandoned the model.

But what we're going to focus on at present is this image of the electron as a tiny particle. Let's first use the diminutive dust particle of our *ordinary world* as a standard of comparison. We can see (and in principle measure) each discrete bit of floating dust as it enters the target area of our shaft of light. Shifting to the quantum world, we can also measure each emitted electron as it makes a discrete target-hit (using Geiger counters instead of our eyes to detect its presence). Okay so far, because each particle is behaving as a particle should, namely as a discrete little bit.

Recall before we proceed further that applying the *Principle Of The Excluded Middle* can become a bit tricky at times, because you do have to be clear on your definitions. With that caveat in mind, let's define what we mean by 'particle'—so we won't make the ex friend's type of error by implying that the electron is something that it actually isn't. We will say that a particle is *a tiny bit of something that has specific dimensions and occupies a specific position at a given time.* Applying this definition to a broader illustration base, picture now the fact that with suitable micro measuring instruments we could accurately measure, say, the size and volume of the tiniest particle of bread left in a certain position on the counter after preparing lunch.

Let's return now to the quantum world. Picture an electron gun firing these tiny particles at a target, with a bevy of Geiger counters set up to measure the spots where the electron particles are hitting. In order to restrict the direction of particle-flow toward the target screen, we will place a barrier between the gun and the screen that allows electrons to move only through a narrow vertical slit. What will form on the screen (seen indirectly through our Geiger counter measurements) is a pattern that's vertically oriented like the slit, though somewhat wider, since some of the electrons are coming out of the slit at a bit of an angle. Because more particles are hitting the target nearer the central area of the vertical pattern (fewer go through the slit at the greater angles), the pattern will be heavier in the central band and a bit lighter near the edges. So far, so good! Same sort of pattern would form if one were squirting, say, particles of water (molecules of H^2O) through a similarly oriented slit.

WAVES

Now we're going to look at the results from another experiment done with pretty much the same setup—but first we need to review the difference between a particle and a wave. Imagine yourself lolling on the sand of your favorite beach, watching the waves roll in. First note that although ocean waves are composed of water particles (molecules of H^2O), the waves are decidedly *not* those particles themselves. Stated alternately, the tiny water particles—as befits particles—have specific dimensions and occupy specific positions at a given time. Each wave, in stark contrast, consists of *progressive motion along a broad front.* The countless water particles that become involved constitute merely the <u>medium</u> through which the wave moves.

Some areas of shoreline, such as the broad inlet on Cape Cod known as Buzzards Bay, result in cross currents that tumble waves together at turbulent angles. Whenever two such waves merge at their peaks, the resulting wave will be much larger in size. By contrast, when the trough of one wave merges with the peak of another wave of equal size, the two will tend to cancel each other out, a phenomenon aptly referred to as 'interference'. This phenomenon constitutes a critical distinction between particles and waves. It explains, for instance, why Newton eventually lost his intellectual battle to establish his theory that light consisted in a flow of particles. The infamous 'double slit' experiment was his undoing.

The original experiment goes back to 1804, at a time when the question of whether light consisted of particles or waves was being hotly disputed. A physician of the time, Thomas Young, projected a stream of light through a narrow slit. As one might expect, a bright and slightly broadened image of the slit appeared on the wall (analogous to the pattern of electrons we've just mentioned). At that point, however, he opened a second narrow slit very close to the first—with a surprising result, for the wall now showed a pattern of equally spaced bright and dark stripes. This was precisely the sort of 'interference pattern' that would be expected if light consisted in a wave phenomenon, where some of the interacting waves from the two different slits would tend alternately to intensify or cancel each other out.

"Wait just a minute," I can hear readers saying, "last time we checked, we were told that light was composed of *particles* called photons that come upon us in discrete units." Well, that's true too, and we'll get to that point in a moment. But I do have to agree right away that if Isaac Newton were alive today, he would consider his particle theory to be vindicated—well, to a point anyway. However, let's return for the nonce to our electron experiment, and pull a 'Thomas Young'. Similar setup as before, but now we open a second slit very close to the first one. This time when we shoot the electrons, the screen will show a pattern of equally spaced bright and dark stripes, an interference pattern that forms the hallmark of wave interaction. Our electrons are now behaving, not like particles but like waves! So which are they *really?* And since the same thing goes for the aforementioned light, which does *it* consist of, *really,* particles or waves?

Physicists can't answer this totally common sense question. Best they can do is to say: "Depends on how you set the experiment up." Sometimes such 'particles' will behave themselves properly and act like the familiar particles of our ordinary world; but under other conditions like the double-slit experiment, these 'whatevers' will act like waves. To keep a sense of humor about such a quantum conundrum, physicists have sometimes used the facetious term *'wavicle'* to designate this oddly acting particle/wave—which term, if taken literally, would be as self contradictory as a 'square circle'.

Our bafflement arises because we automatically—and quite naturally—apply our *Principle Of The Excluded Middle*, an axiom that seems to work flawlessly in our ordinary world. So we end up thinking: The electron (photon, whatever) is either a particle, or it is *not* a particle. And since particles and waves have mutually exclusive attributes, if the electron is a particle, it can't be a non particle like a wave. Something has to give here. And rather than throwing out the consistent results of quantum experiments, it seems preferable to admit that this First Principle, at least as we ordinarily understand it, applies beautifully within the niche to which our species was naturally selected, but bogs down systematically when we try to apply it to the quantum world.[1]

WHAT'S WAVING?

Before ending our analysis of how human cognitive principles tend to break down once we move beyond our ordinary world of mid size objects and their movements, and while we're still on the subject of waves, let's look at still another mind-boggling aspect of that quantum phenomenon. Return to our examination of ocean waves for a moment, and recall that there is a *medium* (water) through which the *wave* is moving. Or watch a football game on TV, and observe a different sort of wave moving around the stadium. Once again, we will find a medium, in this case composed of fans standing and sitting back down. Or speak to a friend, and the sound of your voice will travel via the vibrating medium of air molecules. But try speaking across a vacuum, and your voice will not transmit itself, because there's no medium available to carry the sound waves. Conclusion: wave transmission consists of one or another *vibrating medium*.

So what about electromagnetic radiation (like light, which consists of EMR of certain wave lengths) traveling from the sun to good ole planet earth. In accordance with the above common-sense conclusion, scientists figured that there had to be some *medium* through which the sun's light was moving. So they postulated a very subtle substance and dubbed it the 'ether'. However, when physicists measured the effect of this proposed medium as the earth moved rapidly through it, they found to their surprise that they could provide no evidence for its existence. We were left then with electromagnetic radiation, but with no actual *medium* to do the waving. This amounted to an Alice in Wonderland world, where the Cheshire Cat's smile was still smiling after the cat had gone away.

10

ILLUSIONS, CORRECTABLE AND UNCORRECTABLE

My American Heritage Dictionary defines illusion as "Something that deceives by producing a false or misleading impression of reality." There's no mention of *correctability*, but this notion is usually implied when we speak of illusion. For if our false impressions were not in any way correctable, how would we ever know that our illusions were indeed illusory? That's why, at the beginning of Chapter 1, I noted that "we can be tricked on occasion by, say, visual illusions, but such illusions are correctable. We can also be tricked temporarily by false information, but once again this situation is potentially remediable."

Reminds me of a certain World War II veteran's illusion, going back to the Civil Rights movement of that earlier day. Seems this fellow was freshly returned from his soldiering duties in the desperate struggle to retain our freedom, and he went to his town hall in Alabama to register to vote—he'd been given word that things would be different now that we'd all fought together in the name of democracy. The clerk told him, however, that only those who'd been continuous residents of the town for a number of years were given the voting privilege. So he went off to see his Veteran's agent to produce papers saying that he remained a legal resident of the town, even though he'd been stationed in Europe from before D-Day till war's end. Turned out then, according to the clerk, that only property owners could vote in this particular town. So he went to the Registry of Deeds, returning with proof of home ownership. By this time, the clerk was starting to look a bit desperate…until he remembered that there was also a literacy test that required passing, so he gave the veteran a card inscribed with Chinese characters and asked the prospective voter if he could read what it said. "Yes-siree," replied the veteran, "It says that no Negro's gonna' be voting in this town in 1948." The point of the story, of course, is that experience often enables us to correct our conceptual illusions.

But the term 'illusion' is perhaps most often applied to sensory misperceptions, although here too with an implication of correctability. Not all sensory illusions are of the same nature, nor are they correctable in the same manner. Let me illustrate the simplest type with an instance from my own experience. Gazing from my cottage along the back meadow to the tree line, I thought I saw one of my sheep lying by itself (not a healthy sign in these animals who are innately programmed to stay with the flock). When I looked more closely though, I was able to recognize that the object was actually a dead pine branch with needles the color of my brown African sheep. In this case then, I was able to remedy my visual illusion by means of my visual system itself, simply by attending more closely to what I'd seen.

Many sensory illusions, however, are *not* correctable at the same perceptual level. Gaze along a stretch of railroad, for instance, and the tracks will appear to get closer, the farther one looks. Attend to the scene as carefully as one wants, and the illusion will not change. But if we walk along a railroad bed, we will see that the tracks do not really get closer to each other. Or we could even reason to that fact by watching a train continue to move successfully down the line, a feat that it could not accomplish if the rigid set of its wheel-base had to negotiate progressively narrower tracks. In cases of this sort then, elements from our conceptual system are needed in order to monitor the output of our sensory system, allowing us to stand back from the situation, so to speak, and make the correction.

When it comes to conceptual illusions rather than sensory illusions, one element of our information-processing apparatus is often able to correct an error that's been made by another element of the same system. To illustrate this sort of illusion and its correctability, try the following experiment. Ask some friends what direction they would take, as the crow flies, to get from Los Angeles to Reno. Chances are, they will answer "north and east." That's because they have a rough mental map in which Nevada lies to the east of California; hence, they tend to think that Reno, in Nevada, will lie east of Los Angeles. (They'll add the 'north' part because LA is in southern California, which is mostly south of Nevada). Now hand your friends a globe, and they'll see that their imprecise conception did not take adequate account of the fact that southern California curves substantially eastward—enabling your friends to correct their illusion that LA lies west of Reno.

THE MOTHER OF ALL ILLUSIONS

In Chapter 2, we had asked whether our scientific advances have not amounted to at least "some preliminary albeit partial truths" about the world. On the downside, I would hate to have to relate all the 'knowledge' I accumulated in medical school that turned out to be either wrong or positively misleading due to its incompleteness. I put *knowledge* in scare quotes, by the way, because I don't want to offend philosophers who—at least since the time of Plato—have used the term to mean that what we believe to be the case conforms to what actually is the case.[1] Stated alternately, to speak of "true knowledge" would be redundant, and to speak of "false knowledge" would be an oxymoron. A substantial operational problem arises with this usage, however, because what we *think* we know so often turns out to be wrong, and we don't find out we're in error until sometimes years later (or never).

I referred above to knowledge that was grossly misleading because of its incompleteness, and it's worth emphasizing that our tendency to exaggerate the importance of what we *do* know leads to some of our most egregious errors. That of course is what the famous old story of the elephant and the five blind men was all about.

A STORY ABOUT THE ELEPHANT STORY

The original depiction goes back at least as far as *The Tales of Panchatranta*, a timeworn collection of Buddhist fables. (Our early forbears may have had far less information about the world, but they weren't exactly stupid.) Here's a brief version of the story:

Five blind men, who had felt an elephant at different positions on its body, were asked to describe the animal. The first blind man, having touched its trunk, declared that the elephant was much like a snake. The second, after feeling one of its ears, declared that the elephant was rather like a piece of cloth. The third, who had wrapped his arms around one of the elephant's legs, insisted the elephant was like a tree trunk. The fourth, who had felt along the elephant's massive side, declared that the elephant was like a wall. And the fifth, who had grasped only the elephant's tail, declared that the elephant was like a piece of rope. Listeners immediately recognize of course that each of us is being portrayed metaphorically as a blind man, our limited knowledge leading to various errors in regard to what really is.

But now let's pretend there's an early creature (we'll call it a Romadon), and that various members of this species have the above responses to the elephant already hardwired into their nervous systems. Let's pretend further that Romadons come across elephants in the single context of having to step around these massive animals. Then the representation of an elephant as a "wall" would be good enough to respond adaptively within the Romadon's *ordinary world*, whereas the other impressions would be unhelpful. Those Romadons gifted with the "wall" program would be given a leg-up in negotiating their surroundings—the point being that our imaginary creature need not grasp the 'truth' about elephants. All that's required by the process of Natural Selection is to detect the *pertinent* sort of information required for successful adaptation within an organism's ordinary niche.

For members of *Homo Sapiens*, a similar condition prevails. We do not have access to what-really-is in any absolute sense, but we do have detection and information-processing equipment that provides consistent and useful representations of those aspects of 'What Is' that are crucial for our own adaptation. So what, for instance, if a star that we see in the night sky has long since died, and we are viewing only rippling waves of light that this now dead beacon had emitted millions of years earlier. Light transmission, as far as our built-in computational systems are concerned, seems to take no time. And such indeed *is* the case, for practical purposes, within our *ordinary world* of limited distances. To say that this computational strategy of our nervous systems involves "an illusion" is of course technically correct; nevertheless, it's a stable illusion that happily simplifies the task of adapting to our surroundings.

Another way of emphasizing this point is to say: Cognitive systems were never selected to gratify the philosopher's longing for 'Truth'. Rather, these systems have been selected for their utility in promoting adaptation to the key features of a particular organism's ordinary world, our own included.

UNCORRECTABLE ILLUSIONS

Correcting what *may* be a visual illusion as we fix on a given star bears some resemblance to the situation involved in the earlier train-track illusion, since both cases involve sensory illusions that are uncorrectable at the sensory level. Nevertheless, there's a considerable difference between the two cases. That's because all normal members of our species can correct the train-track illusion directly, using no more than their everyday conceptual equipment. However, when it comes to correcting our certainty regarding the distant star's continuing existence, we must bring the whole apparatus of modern astronomy into play—refined sense extenders in the form of powerful telescopes, as well as highly technical conceptual models that make extensive use of advanced mathematics. Most of us, therefore, are reduced to accepting the consensus judgment of experts in the area.

But there's an even more crucial difference. Although we accept the judgment that a given star's continued existence *may* be an illusion, not even the experts can tell us what has actually happened to that celestial sphere in the millions of years elapsed since the information we are currently receiving was originally sent on its way. That task would be equivalent to surmising the present status of the person who sent the message-in-a-bottle that's just now washed onto shore. So we end up with a lesser correction, not knowing whether the given star is still present, but only that it might not be. If, during the coming year, information reaches us that the star has exploded into a supernova, we will of course have additional information helping us to judge that the mere possibility of its non existence has turned into a reality.

It still follows of course that *if* our illusions were not in any way correctable, we could never know they were illusory. So an illusion we describe as "uncorrectable" must be at least partially correctable. We have just described such a partially correctable perceptual illusion; and in the previous chapter, we reviewed some instances of conceptual illusions that were only partially correctable. For instance, we have a robust illusion that our First Principles apply everywhere and at all times as they do in our ordinary world of mid size objects and their movements, so we feel almost vertiginous, intellectually speaking, when the application of these principles doesn't square with findings from the extra ordinary world of, say, quantum mechanics.

Is the electron a particle, OR is it a wave? It certainly can't be both—we have the incessant urge to claim—because particles and waves have incompatible properties. We can eventually correct our initial illusion by inferring that our First Principles, as we customarily apply them, must be 'special case' laws adapted to our ordinary world where they seem to work so flawlessly. But we're still not able to make a truly positive clarification about what really gives rise to this 'wavicle', so our correction is only partial.

But what about someday? Perhaps *someday* we will be able to conceptualize clearly what's involved in the 'whatever' that we refer to as the wavicle. This leads us to a consideration of what we might refer to as *The Principle of Cognitive Limitation*.

PRINCIPLE OF COGNITIVE LIMITATION

We saw in Chapter 1 that the deerfly is able to detect vertical-end-stopped objects and to respond vigorously to such objects in motion. But it has no computational program enabling it to make a distinction between edible and inedible objects of this nature. Interestingly, such an additional computational program might not seem that hard to devise. For instance, we could imagine a functional algorithm that went like this: If your proboscis hits something too hard to penetrate, immediately cancel the eating maneuver and take flight again. But for that strategy to be useful would require additional programs to identify and maintain in memory this particular object. Otherwise, the deerfly might take flight and then land again on the same object, over and over and over. So given the limited computational space available to the deerfly, it may well have hit on one of the best engineering compromises available. Nevertheless, the result is that the present day deerfly is *never* going to be able to distinguish food from non food.

And we can see analogous limitations in animals possessed of much more complex nervous systems. Present laboratory rats with a maze-running task, for instance, and they'll do well. Now set the rats up with an algebraic problem, and they won't have a clue. Not simply that they can't solve the problem today. Even if we were to undertake a long and painstaking educational effort, rats still wouldn't have a clue. That's because members of this species lack the computational apparatus necessary to handle such problems, so they're *never* going to be able to understand what's involved.

How about us then? Are we the sole species exempt from the *Principle Of Cognitive Limitation?* I think only this sort of extreme stance would tempt one to say "Well, we don't know what's actually involved in a wavicle at the present time, but *someday* we undoubtedly will." The problem is that human cognition operates in full accord with our *Principle Of The Excluded Middle.* Our cognitive systems do not seem to have available an additional Program X that might enable us to understand what the wavicle is inherently about.

Since our cognitive apparatus operates along a series of levels, one aspect of our conceptual system can monitor another aspect in operation. This frequently enables us to attain at least some faint glimmerings of that which seems to lie beyond our proper ken—a point that we'll examine more closely in the upcoming chapter.

11

SEEING FURTHER THAN WE CAN SEE

As a young boy, I once read an illustrated popularization of the Special Theory Of Relativity. I can recall feeling unhappy with the various cartoon depictions, however, because none of them really displayed to me Einstein's brave new world of four dimensions. Afterwards, I almost broke my mind trying to put all four of these dimensions together into one neat picture. Any reader who's ever wasted time trying to accomplish the same feat won't be surprised to hear that I didn't have much luck. Problem is that our computational equipment obligatorily breaks the space-time continuum into mid size objects that change over time, so members of *Homo Sapiens* cannot directly depict all four dimensions together.

Not so fast, readers might say. Isn't that exactly what Einstein accomplished? And if he could lead the way, why can't we follow? But Einstein wasn't really able to *visualize* his four dimensions together, any more than we can. His feat involved a conceptualization made possible by the tools of advanced mathematics, enabling him to correct the illusion of most human beings—including even the great Sir Isaac Newton—that space has three absolute dimensions, and that objects within these three spatial dimensions interact across the separate category of time. (Interestingly, this case involves a human sensory illusion that was strongly reinforced by common-sense conceptual analysis.)

UNCORRECTABLE CONCEPTUAL ILLUSIONS?

Nevertheless, even in this case, human beings were eventually able to correct the illusion. So might we not argue that our human conceptual apparatus, at its best, is capable of correcting *all* of our illusions? Might we not contend, for instance, that current conceptual analysis of astronomical distances and the speed of light has also allowed full correction of our perceptual illusion as we gaze starward? For what sense would it make to apply the word 'illusion' every time we lack certainty about all the features of a given situation? If we were to follow such a procedure, we'd end up declaring that *everything* we experienced involved an illusion.

This argument is attractive, because it employs a no-nonsense approach to the use of our human cognitive capacities. Nonetheless, the fact remains that the Elephant Story really *does* apply to any and all of our experiences. That is, since our knowledge is always limited, and since limited knowledge inexorably involves varying degrees of distortion, there's an illusional aspect—of this kind at the very least—when it comes to any and all of our human experiences.

We should quickly reaffirm, however, that stable illusions within any organism's ordinary world, our own included, can amount to highly useful engineering shortcuts. Furthermore, those illusions actually selected usually interfere not a wit with successful adaptation. In our case, for instance, systematic sensory failure to recognize that the star we are seeing may no longer exist makes no practical difference. And the eight minutes that light takes to travel from our nearby sun makes no practical difference either. That's because, in our *ordinary world,* distances are so minuscule—relatively speaking—that our systems do well to register the transfer of light as instantaneous.

CONTINUUM

One of the most intriguing limitations of our computational apparatus derives from its compulsory use of objects-and-their-movements. This involves sensory objects like the *things* we see, as well as conceptual objects that constitute the *items* of our mental activity. I've italicized the words 'items' and 'things', by the way, in order to emphasize the point that I can't escape from thinking in terms of objects; for these alternate words are, after all, no more than synonyms (and I might as well throw in a fancier one like *entity).* The human conceptual apparatus, as it turns out, has made no provision for us to grasp a continuity uninhabited by objects.

Let me now point, however, in the general direction of this feat—albeit a feat that's actually beyond our abilities—by using gradual color change for purposes of illustration. Imagine a large vat of blue pigment, into which we slowly stir a bright yellow pigment. Gradually, the color will turn green, but there's no exact point at which this transformation will have occurred. The change will seem to be *continuous,* although we can conceptualize in the direction of this continuum by attributing color names to various segments along the way. For instance, when enough yellow has been added to create early stirrings of a greenish blue, we can refer to that color interval as 'turquoise'; and further along in the process, when we've reached a green with still discernible traces of blue, we can call that color 'aqua'. And no doubt the interior decorators and paint merchants among us could add a bevy of additional names to designate more subtle changes of shade along the way. But if we were to end up with more than a hundred such designer labels to register different intervals of change, it would illustrate all the more forcefully that the only way our conceptual systems can deal with a continuum is to *digitize* it; that is, to break it up into discrete units of smaller and smaller size.

How about our visual system though? Were we not able just now to view a continuum of color from blue to green? Well, we certainly did get closer to such a direct experience—enough to provide us with glimmerings of what a continuum might be like. But actually our color detection systems work in stepwise fashion also. For example, we're unable to pick up small differences in wavelength, even when our sense extenders (*i.e.* color meters) assure us that such differences are indeed present. So our visual sensory systems are also reacting stage-by-stage, even on those occasions when we have the illusion of continuous color change.

BUT MAYBE THE WORLD <u>IS</u> DISCONTINUOUS

We could well argue—and for all I know it may be so—that the reason our detector systems and our information-processing systems digitize the world is because that's the way the world actually is. After all, not only does our *ordinary world* consist of more or less discrete objects and their movements, even the quantum world appears digitized into discrete quantities. For instance, instead of electrons varying in energy along a continuum, these minute particles jump from one discrete energy level to another. Same in the case of light, which arrives in discrete packets of energy. The photon then is not merely a conceptual object we are forced into using by a cognitive system that can employ only objects. The word 'photon' refers to an energy packet that exists in nature whether or not we think about its existence.

Nevertheless, we might still argue that as submicroscopic as the quantum world indeed is, it's still far from rock bottom in terms of reality, as witnessed by the work of string theorists in recent years who have been attempting to provide even more basic representations of 'What Is'. And while their efforts boggle our minds to an even greater degree (using eleven dimensions and more, for instance, instead of a mere four), these theorists still obligatorily employ discrete units in the form of one dimensional strings in order to erect their theories. Bottom line:

<u>Even if all-that-is did consist of one glorious field of force varying markedly in intensity, we would never be able to directly grasp that reality.</u>

WHERE IS THE ELECTRON?

And even within the quantum level of reality, applicability of our notion of discrete objects has its inherent limitations, a point we might illustrate by reviewing an explanation that's been commonly provided for why Heisenberg's *Uncertainty Principle* applies in this domain. The question is: How are we to explain, say, that we can never get an accurate measurement of an electron's position *and* its momentum at the same time? Here's the explanatory story: Think of the electron as an extremely light billiard ball (hence, having highly specific dimensions). We need to use a probe in order to ascertain its exact location. But whatever probe we use, the electron is so light that it'll be bumped off course a bit by the measuring probe. So we may end up knowing where the particle is at that moment, but our very measurement procedure has changed its momentum to some extent.

This explanatory story squares totally with what we'd expect in our ordinary "billiard ball" world. Hence, it has the virtue of being common sensical. That, ironically, is the first reason we should probably be suspicious of the story's total accuracy—since the quantum world seems bent on thumbing its nose at our ordinary ways of thinking. So once we start to imagine the electron as a teensy light-weight billiard ball, we should already start worrying that we're off in the wrong direction. As we saw in Chapter 4, particle physicists use the word 'particle' differently than we do in everyday life, where the word implies an object of discrete dimensions, like a dust particle or a particle of food. The electron 'wavicle' will never fit, even in principle, the category of discrete object with discrete dimensions occupying a precise spatial position. Hence, this quantum particle is, in some respects, not fully digitizable.

THE GHOST OF A DEPARTED QUANTITY

The human conceptual system, even in its most sophisticated mode—as for instance when we use abstract mathematical reasoning—cannot transcend our obligatory reliance on objects-and-their-movements. Let me illustrate: Suppose I drive the 40 miles from Boston to Worcester, and it takes me an hour to make the trip. How fast was I going? Easy question to answer: just divide the distance by the time needed to traverse it. So my average speed (velocity) during the trip turns out to be 40 miles per hour.

But now let me share with you something about my trip. The venerable Worcester Turnpike, a state of the art four-lane highway when it was constructed in the early nineteen-hundreds, has been overwhelmed by the volume of traffic that circulates ever more sluggishly through our burgeoning suburbs. With the addition of strip malls galore, not to mention complexly arrowed lights to bring all traffic to a periodic halt (if not totally to its knees), there were but scant moments when my pickup was actually traveling at a velocity of 40 mph. And most of those moments involved times when I was either speeding up or slowing down—between the modest stretches when I was able to cruise at 60 mph (the speed limit of 55 mph, plus the extra five granted by the State Police according to gentlemen's agreement).

So let's begin with me stopped at a red light. The control turns green, and I accelerate my vehicle gradually from 0 mph to 60 mph. Along the way, I will at one point in time be going 10 mph, at another point be going 20 mph, at a still later point 30 mph, and so on, until I level off at full speed. One way of expressing my acceleration would be to measure the time it took to go from a full stop to my steady state speed, say, 10 seconds from 0 to 60 (though automobile advertisers usually save this sort of figure for cars that go "from zero to 60 in four seconds"). Another common way to portray my rate of acceleration would be to give my average rate of acceleration for each second, a figure obtained by dividing the 60 mph *change in velocity* by the 10 seconds it took to accomplish that change. Expressed in this way, my average acceleration was 6 mph per second.

Notice that when I first mentioned the time my trip took, I referred to my "*average* speed" (or velocity). And when I mentioned acceleration just now, I referred to my "*average* acceleration" over a period of time. So I was not really addressing my earlier statement that "*at one point in time*" I was going 10 mph. Reason is that while we tend to think of a continuous acceleration, traveling at such and such speeds at particular points in time—in this case, 6 mph at exactly 1 second, 12 mph at exactly 2 seconds, and so on—we can't actually measure speed (velocity) except as distance over elapsed time, or acceleration except as change in velocity over elapsed time. So if no time has elapsed (and no time can elapse, by definition, at any specific *point* in time), we cannot really speak of velocity.

Reason I take the time to remind you of this fact is to note the difference between a *continuous* acceleration and the only way our cognitive systems allow us to represent that sort of continuum, namely by averaging over small units of elapsed time. Even sophisticated mathematical thinking must digitize a continuum into *units* (which are of course types of objects). But suppose that for a given purpose, average acceleration *per second* is not finely honed enough for our needs.

We could instead measure changes over thousandths of a second, or if our measuring devices were good enough, millionths of a second.

Mathematicians, who unlike physicists don't have to do actual measurements, have pushed this division of smaller and smaller units to the limit, thereby providing us with an *almost-continuity*. Here's how they do it: Conceive, they say, a number that though greater than 0 is smaller than any number you might imagine. That sort of vanishingly small number was crucial in order for Newton and Leibniz to develop their Calculus, and it's referred to as an "infinitesimal"—which drew from Bishop Berkeley the sardonic remark that they were dealing with the ghost of a departed number. And even today, if one Googles onto "infinitesimals," one can witness mathematicians continuing to fuss with each other over the exact significance of this ghost. My present point, however, is simply that our cognitive equipment, even at its mathematically most sophisticated level, can deal with the notion of continuum only by digitizing it, that is, only by representing a continuum as objects-and-their-movements.

And notice that as I've been talking about a continuum just now, I've been able to do so only by representing *it* (sic) in the form of a word, namely 'continuum'. Given my human information-processing apparatus, I have no other option. But 'continuum' then ends up as simply one more object, albeit a verbal one. We'll take a closer look in the upcoming chapter at the straitjacket of sorts that this representational method places on our conceptual/language apparatus.

12

WORDS ARE LIKE BOXES

It's true enough that we often use the word 'continuous', though as we noted in Chapter 11, even with the sophisticated conceptual tools available to us in mathematics, we always end up digitizing stuff. We have no choice because our information-processing systems work in the obligatory mode of objects-and-their-movements. This is the case of course with words as well as numbers. In fact, the flexibility of words (also known as lack of precision) is often so pronounced, it can at times suggest the illusion that our systems are operating directly by way of continuums.

Suppose, for instance, I try to describe to you what the land on my homestead is like. I might say that it's part woods, part meadow, part swamp, with an area of lawn immediately around the cottage. If you were to ask what portion of the land fell into each category, I'd have to answer: "It's hard to say exactly, because one type of area often melds so gradually into another that I'm not able to draw exact borders." Sounds then as if my words are depicting a continuum of change.

But if I were pressed for more accurate representation of my land, I would have to admit that "it's about half woodland, more than a quarter meadow, less than a quarter swamp, with just a small rim of lawn immediately around my cottage." In other words, when I move to verbal containers that are a bit less vague, it becomes clearer that I'm digitizing information all along—just that my original containers were imprecise enough to create an illusion that I was cogitating directly in a continuous mode.

What would I do if pinned down even more concerning, say, the gradual change from woods to meadowland? One choice would be to introduce the notion of a 'transitional zone'. That would amount, of course, to the introduction of still one more *thing* (woods > transition zone > meadow). Or I might provide an operational definition of woods and meadow that would enable me to make a sharp distinction (relatively speaking). For instance, I might define meadow as a region that's almost totally grass, and that any tree in the area must be at least 30 feet distant from the others. Otherwise, it's still woodland. Point is that the illusion of continuous change produced by my original statement that "one type of area often melds so gradually into another" comes from the vague borders of my original verbal containers, not from my processing the data as a continuum.

Furthermore, shifting from the conceptual to the perceptual level doesn't help much. For while I do at first seem to see a continuous change from woods to meadow, that's because my initial impressions don't focus on details. But when I start looking more closely at the "continuous change," I can see that my visual system registers it also in stepwise fashion.

THE FLEXIBILITY OF WORDS

The more we focus on the *flexibility* of our words, the more we can become frustrated with the downside of this phenomenon, namely, the oftentimes frustrating *lack of precision* that goes with the territory. It's as if we've placed an order that should come in a standard-size box, but the box turns out to be so variable in size that more items or fewer items get packed into recurrent orders. And worse still, not only is the box likely to be of different size the next time we order it up, some items we thought we were ordering might not even get packed inside the next time around.

But of course, if we don't focus too much on this downside, we can better appreciate the upside. Given verbal containers that are more flexible in size and content, we can get by with many fewer boxes, which amounts to a highly effective engineering shortcut. We achieve a huge savings in the amount of vocabulary our systems have to carry around; furthermore, the savings generally work out well, since *approximation in life is the name of the game.*

And consider an additional advantage: We have a number of people among us each generation whose cognitive proclivities would almost certainly lead them into a life of crime. But by channeling their antisocial traits in a somewhat less destructive direction, we end up with lawyers, word-spinners who can make an almost honest living from reinterpreting what their clients truly and actually meant by the words they originally inserted into the contracts that are now being contested by other, obviously mean-spirited, signees (who in turn will hire additional lawyers, keeping even more of these potential muggers blessedly off the streets).

TECHNICAL DEFINITIONS

It's easy of course to trash lawyers, because we all know what nefarious villains they are, with the exception of my brother. But in fact almost all professionals are called upon to sharpen the borders of the words they use in order to fashion these objects into more dependable and less ambiguous containers. We could illustrate that point from my own field of psychiatry:

Suppose one of your friends says "I'm really feeling down today." Since "feeling down" is pretty much the same as "feeling depressed," does that mean your friend is now suffering from a depression? And if so, shouldn't he be given Prozac or something? And while we're at it, maybe some psychotherapy? Here's how psychiatrists would respond to that question: First, they'd make a distinction between the word 'depression' as a symptom and the word 'depression' as a diagnosis. Same verbal box, different contents. "Feeling down," "feeling blue," "feeling awful discouraged," and "in the dumps," are roughly translatable into the symptom-of-depression. But that's far from identical with the clinical condition called depression. Psychiatrists require a whole set of additional items, a certain number but not all of which are necessary in order to make a diagnosis (e.g. minimum 2 weeks feeling depressed, sleep disorder, appetite disorder, exaggerated pessimism, irrational feelings of guilt, lack of energy, lack of usual feelings of pleasure, social withdrawal, and so on).

We could refer to the above as an operational definition, for as in the earlier case of my woodland and meadow, it permits a sharper delineation of what's being referred to—what items are to be packed in the box—and does so in a way that's readily discernible. This particular type of operational definition also amounts to a technical definition. Why? Because it's not just a working definition that some individual is using informally. It's been codified by a group of people who agree to communicate among themselves with this meaning in mind.

Such terms are useful for a number of reasons. For instance, if physicians use an APA diagnosis[1] like "Major Depressive Disorder," that little three-word phrase will communicate to any professional who reads the report entire paragraphs of information. Furthermore, a patient who fits the multiple-item profile involved will be more likely to respond to interventions that have been found useful in the treatment of this condition.

BUT THE TOOLS FOR FURTHER DEFINITION ARE STILL WORDS

While technical definitions *do* delineate the edges and contents of our verbal containers to an extent that often proves useful for a given endeavor, the tools-for-sharpening consist of still more words. We have no choice. Yet each new word we introduce will itself contain a certain amount of ambiguity, which in turn will require further effort at definition, hence still more words. So we're never able to reach the rainbow's end of excluding all vagueness. Let's stay with the technical definition of depression a moment longer in order to illustrate the point.

One of the commonest symptoms of depression is, not surprisingly, feeling depressed. But now suppose someone asks me what I mean by "feeling depressed." When someone tries to pin me down on some basic notion like that, I usually find myself starting my answer with "Well, you know, it's when...." The well-you-know part amounts to "Heck, you know what I mean, or at least you darn well should..." That approach can sometimes put the other person off from landing too hard on the vagueness of the verbal containers I'm using. In actual clinical practice, however, the point amounts to far more than mere nit-picking. Let me illustrate by describing a sequence that's come up repeatedly in 45 years of clinical practice:

Let's say a patient looks depressed, so I ask: "You been feeling depressed lately?" to which he may shake his head and say "No."

"Been feeling blue?"

"No."

"Dejected?"

"No."

"Sad?"

"No."

"Down in the dumps?"

"Yeh, doc, I've been feelin' awful down in the dumps these days."

Now I've never been able to figure out why a patient sometimes accepts one synonym for "feeling depressed" and rejects another one, but it's not rare for that to happen. You can also see how I handle it, namely by translation: Feeling down in the dumps = feeling depressed.

And sometimes the plot thickens even more. Sometimes a patient will answer: "No, I'm not depressed; just that I've been awful worried; I don't think things are ever going to work out right again." Interestingly, the APA Diagnostic Manual tends to put "excessive worry" into the carton of Anxiety Disorders. (I say 'carton' because a number of diagnostic boxes are squished into this jumbo container.) Often, however, "excessive worry" turns out, as in the above instance, to be a prime way of expressing excessive pessimism, which in turn forms a key facet of feeling depressed. In fact, a whole technique of psychotherapy has been developed that focuses on the clarification of this depressive way of thinking. So, should we put "excessive worry" into the box of anxiety, or should we see it as a so-called "depressive equivalent?"

We're not even going to try to resolve the issue here. The present point is simply this: While technical definitions do help to make our verbal containers more precise, the help is only partial. That's because we achieve increased precision in such cases only by the use of still more words, and each new word we introduce brings with it additional fuzzy borders. In our illustration, the clinical diagnosis of depression was sharpened by specifying what boxes are to be placed therein (and what other boxes, by the way, are to be left outside). Then we examined one of the enclosed boxes, "feeling depressed," and saw that its borders are not all that precise either. In the end, we're forced into the straitjacket fashioned inexorably by the verbal containers we use—containers that make only an approximate fit for what we're trying to get at.

PROCRUSTEAN BED

Our verbal boxes are few, relatively speaking, but the real-world instances to be matched up with these conceptual containers provide unending variations around any given theme (almost as if the real world really is a continuum). So how are we to make the best fit? The ancient Greeks were in the habit of addressing such issues by story telling, and I've always liked the one about Procrustes. He, as it turns out, was an innkeeper, though he must have run a very modest enterprise because he had only one bed—and that of none too ample proportions. So how would he handle the situation when a customer came along who didn't quite fit the bed that was available? No problem. Procrustes would just cut the person's legs off to make a satisfactory fit. Fortunately, we don't need to inflict such violence on the other people in our lives, but we are called upon quite regularly to cut and twist the things we come across until they squeeze into one or another verbal container we happen to have available.

ESSENCES AND ESSENTIAL DEFINITIONS

In an effort to fashion our trashbag containers into sharply contoured verbal vessels, philosophers used to be especially fond of pursuing the following endeavor. Each thing, they claimed, possesses an *essence*—that which makes it what it is—plus a whole bunch of additional stuff that's merely accidental. To illustrate: Suppose I have my current locks shorn in skinhead fashion. I will have changed, but even though I look different, I'll still be the *same* person. My hair style, in philosophical parlance, is an "accident" (though in the above case, it might also be a tragedy). So, what then is my essence?

The essence of a human being, according to the Scholastics (who were heavy into essentialism) is *rational animal.* In arriving at this definition, they followed Aristotle's two-part approach of first naming a bigger box (called the 'genus') and then inserting a smaller box (called the 'species') within this bigger box. In the present instance, of course, the bigger box is 'animal', and the smaller box is 'rational'. We'll gloss over the fact that our notion of what constitutes an animal is rather fuzzy at the edges, and focus instead on the obvious difficulty that a whole bunch of human beings don't qualify as rational (young babies, for instance, or the severely retarded, or the lately demented).

No problem, the Scholastics would say. Young babies *will* grow into rational adults; the severely retarded *would have* become rational adults were it not for intervening pathology; and senile seniors *were* rational before amyloid plaques and neurofibrillary tangles got their brains. Note then that Scholastics started with an essence they tried to capture within the <u>static containers</u> of our words/ concepts, but were soon required to move in the direction of <u>evolving process</u> (yet another static container, albeit one that's pointing in the direction of something our computational systems just aren't up to). And that becomes evident even before we ask questions like "At what exact *point* does the growing child become rational?" Once again, it almost seems as though there are continuous processes (one grand and gloriously overarching continuum?), but that the best we can do is to assemble a bunch of sequential snapshots to suggest this seamless continuity.

THE ESSENCE OF 'GAME'

For those who enjoy word play but who've become bored with scrabble, I suggest that you read Wittgenstein's entertaining comments in his *Philosophical Investigations*[2] as to what in effect might constitute the essence of a 'game'. He noted that we apply this word to a whole bunch of different things, none of which need have even one basic feature in common. He ultimately concluded that whenever we apply the word 'game', we're talking about some item that bears a "family resemblance" to the other items that we call games. In similar fashion, the many patients suffering from 'clinical depression', though often varying greatly in their clinical presentations, will all show a family resemblance to one another. (Please don't ask me though for a technical definition of "family resemblance." You know, it's when...)

ESSENTIALIST THINKING IS NOT JUST IDLE

Lest I leave readers with an impression that the issue of essences is no more than philosophical fun and...*uh*...games, let me remind you that real, live, ongoing societies are suffering painful splits involving application of such static conceptual boxes. Indeed the proposed essence of a human being provides an apt illustration. For at what point in development does a fertilized egg become *essentially* a human being, and hence, subject to the same rights as any other human being?

Right-to-lifers judge that the essence of a human being is established at the moment of conception; hence the fertilized egg is a [human being]. If so, then any planned abortion amounts to the volitional killing of an actual human being. Pro-choicers judge that an embryo is only an evolving precursor to the separately existing [human being] who would eventually be born; hence the pregnant woman's judgment should be weighed heavily into the decision about continuing the process toward full humanhood.

This is not the place to even attempt a mediation of this particularly divisive issue. The present point is simply that our obligatory dependence on static conceptual boxes, along with a tendency to fix the 'essential' ingredients placed in a given box, can easily lead to the assuming of polar positions that leave little room for negotiation. It's not always just fun and games.

Part II
SELECTING ON THE BASIS OF VALUES

INTRODUCTION TO PART II

Thus far, we have emphasized the fact that organisms can do no more than sample the information overflowing from their environment. Animals fashion the relatively scant information they're able to detect into representations of their surroundings—*approximations* that are good enough to allow survival and reproduction within a given species' ordinary world.

The information-processing equipment of our own grand and glorious species provides us with sophisticated representations of ourselves and our environment that far outstrip anything given to the other animals on planet earth. Nonetheless, the same principle ultimately applies. We detect only certain things, and we process the information that we're able to glean in only certain ways. Specifically, we've been naturally selected for adaptation to a world of "mid-size-objects-and-their-movements." We cannot <u>not</u> experience the world in this manner because it's a core element of our information-processing programs. And though sophisticated sense-extenders (chemical assays, microscopes, radiation detectors, and on and on) have greatly increased the extent of information available, our processing systems still limit us to computations over *objects-and-their-movements;* and *that* continues to be the case even when using our conceptual equipment in its most advanced mode, as when we perform abstract mathematical manipulations.

BURIDAN'S ASS

We will now move on to an additional facet of our information-processing apparatus, one that's absolutely crucial if we're to respond adaptively to the complexities of our world. To illustrate the problem in need of solution, let me relate the story of Buridan's ass. The name of the tale comes from its supposed author, a 14th century Scholastic, Jean Buridan (though probably better attributed to one of the philosopher's critics). At any rate, it seems there was this ass who one day found himself exactly midway between two scrumptious bales of hay. Being strongly but equally attracted in both directions at the same time, his urge to move one way was precisely balanced by his urge to move the other way. Tragically, this led to his starving to death in the midst of plenty.

None of my sheep, I should hasten to reassure readers, have ever run into this problem when I've set out winter hay on the snow covered ground. Could be they're smart enough never to put themselves in that position, but it may also be that the natural dynamics of nervous systems tend to keep connections cycling so as not to get caught at a totally static set-point. (Unfortunately, some human beings of the Hamlet type *do* on occasion get themselves stuck darn near such a set-point, and become prone to paralyzing degrees of indecision.)

WEIGHTING

The general solution to this kind of problem involves adaptive *weighting* of opposed behavioral programs—weighting that also varies under different circumstances. For instance, the urge to stop at a water bubbler will vary, depending on our degree of thirst. We will plan, as in Part I, to introduce the subject by way of less sophisticated organisms. Since factors underlying their behavior are generally less complex, it's often easier to get a handle on what's involved.

VALUES

When dealing with the weighting apparatus apparent in our most sophisticated systems, we usually employ the fancier word 'values' rather than a more workaday term like 'weighting'. The principle remains the same, however, so we might aptly say that our application of values can be traced back to weighting mechanisms that evolved in less complex organisms over preceding eons.

We will focus on our most crucial value system, one that's absolutely necessary if an extraordinarily complex social species like *Homo Sapiens* is to make it through even a single reproductive cycle: namely our *sense-of-fairness* program. And we will see that this program bears interesting comparison to our language program. We have already reviewed the fact that the latter device is built into our species, beginning to unfold spontaneously during the second year of life. Yet, at the same time it's clear the particular language learned will depend on that spoken by one's familiars.

Our *sense-of-fairness* program also unfolds spontaneously, its rudiments being observable during the same early period. And similarly, the particular 'language-of-fairness' that's learned will also depend on that displayed by one's familiars (albeit in more convoluted fashion). As regularly happens in the case of complex behaviors then, we can see the reverberating interaction of nature *&* nurture—the oversimplified dichotomy of nature *or* nurture never really filling the bill.

What may be new to some readers is the notion that our human ethical sense (basically our sense-of-fairness program) can be accounted for within the context of Natural Selection. At a proximate level of explanation, no hypotheses beyond the ordinary workings of nature are required.

13

HOW DO ANIMALS MAKE CHOICES?

I always liked Albert Schweitzer's notion of "reverence for life," but I've found that my fondness for the idea doesn't always translate into action. For instance, when a deerfly lands on me, I whap at it, fast as I can. Their bites hurt! Deerflies are not as agile as houseflies, no doubt because deer and related animals don't have slapping hands to 'select out' less nimble members of the species. Nevertheless, these critters do take evasive action in response to a potential victim's forceful protests.

Now I suppose at this point my mentally nimble readers may be complaining: "The rascal's contradicting himself; he told us a while back that deerflies are programmed to *approach* moving objects!" And indeed I did, so we have to give the critters credit for an additional behavioral program that goes 'Take flight if something starts to move very rapidly toward you'. Note that these two programs call for totally opposite reactions: One directs approach-behavior in response to moving objects, but the other calls for evasive action in response to rapidly moving objects. Furthermore, if the 'evasive program' is to work successfully, its implementation has to be given heavier *weighting* than the 'approach program'. Whenever a close call arises, that is, the evasive program has to take priority. But before pursuing our investigation of such weighting contrivances, it may be helpful to take a brief additional look at how Natural Selection comes into play in the behavioral arena.

NATURAL SELECTION OF BEHAVIORAL PROGRAMS

It's no coincidence that Darwin initially formulated his thesis of Natural Selection heavily on the basis of anatomical traits. One brief (albeit enlightened) observation of a Galapagos finch, for instance, will disclose the shape of its beak, and it's a very short stretch from there to an inference about the optimal use for this kind of tool: Long slender beaks will be more suitable for exploiting certain sources of food, short heavy beaks more suitable for exploiting other sources, and so on. Nevertheless, the best anatomical equipment in the world won't get the job done if left to its own devices. A large variety of behavioral programs also need to be on line. However, since the behavior involved may take significant time to unfold, extended periods of observation are generally required in order to elucidate the latter programs.

Let me illustrate the point from my own sheep. Newborns need their vital anatomical equipment, of course. But in order to survive, lambs also need not only milk but a variety of behavioral responses from their mothers. Without such 'maternal care' programs, all lambs would be doomed, a point that's been made graphically (and to my horror) on those thankfully uncommon occasions when maternal behavior doesn't kick in. More on that later, but the present point is simply that we tend to take behavioral programs for granted until they're not working. Only then do we appreciate that the requisite behaviors don't just come out of nowhere; subtle as they are, they're part of the necessary equipment of all organisms. Think of this analogy: I see my computer screen and keyboard in plain sight, but I don't see my word processing program; yet if that program were not installed and functioning, there's no way I could be using my computer to write these lines.

As in the case of anatomical features, information for behavioral programs is passed along within an animal's genetic code. Most coding changes (*aka* transcription errors) will lead to diminished operational ability, and animals with such diminished ability will tend over time to be selected out. Sometimes, albeit rarely, such a change will lead to enhanced operation. In such cases, the organism so gifted will be more likely to survive and propagate, tending to make its kind gradually more prevalent. Natural Selection, that is, applies to behavioral programs just as forcefully as it applies to more readily observable physical characteristics.

PRESET PROGRAMS vs. PROGRAMMABLE PROGRAMS

Our personal computers seem able to perform miracles when armed with the proper programs, and the same applies to organisms. In both cases, built-in programs may be totally prewired to operate rigidly (like our simple hand-held calculators); however, highly complex computers can be programmed to learn from experience.

Insects have been invaluable when it comes to sorting out the difference between the behavioral output of rigid programs *vs.* output due to complex programs requiring extensive feedback before becoming functional. That's because most insects don't hang around this vale of tears long enough to learn the ropes. They have to be up and running from the start, without any opportunity to learn at their mother's knee.

Nevertheless, results can still be quite spectacular. One of the more famous examples comes from observation of a variety of Digger Wasp (the Sphex). Without opportunity to learn from experience, the female of this group will sting a suitable insect correctly and lug it to the entrance of a burrow that she's previously prepared. Next, the wasp enters the burrow to check things out, returns to the surface, drags her victim down the hole, then lays her eggs on its body within this safe enclosure where the developing embryos will be protected and nourished.

However, when entomologist, Jean Fabre, moved a paralyzed cricket mere inches away from the burrow's opening while the Sphex was checking inside, the wasp repeated the whole procedure from the start—dragging her prey to the opening and rechecking the burrow again—for the forty repetitions that Fabre had the patience to document.[1] Examples like this make clear that quite complex behavioral programs can be loaded into even tiny nervous systems, but that such preset programs tend to unfold rigidly once they're set in motion.

The case is different with more sophisticated animals. Take the cotton-tail hanging out recently near my front gate. He munched on the meadow grass to his heart's content, but he wasn't keen at the beginning when I'd come along. First he'd freeze; then as I drew nearer, he'd take off like a bat out of hell. These totally opposite behaviors are preprogrammed from Peter rabbit on down, and they involve *differential weighting*. The 'freeze program' gets heavier weighting when the danger is more distant; but as the danger moves closer, the weighting shifts in favor of the 'flight program'. The weighting, however, is also subject to modification from experience. So when this particular rabbit became more familiar with me (and once he noted from my name tag that I was not farmer McGregor), he'd allow me to get progressively closer before taking off.

MATERNAL CARE PROGRAMS

Maternal care programs are crucial in all mammals, though I suppose we might even refer to the Sphex's preparations for her offspring as a primitive program of this kind. But while it's true that her provisions are impressive, the Sphex doesn't have to deal with the complex behaviors that would arise when her fertilized eggs turn into full-fledged offspring. With mammals, the case is different. Even though my ewes, for example, show some maternal behaviors that seem pretty mechanical (e.g. the licking off of their newborns and the assumption of bodily posture ideal for suckling), things get lots more complicated.

To begin at the beginning, since ewes have only milk enough for their own lamb (and barely enough for twins on occasion of their birth), the mothers need to be able to distinguish their own offspring from others. That's a skill we tend to take for granted—as if it just magically happens—until the program doesn't work properly and the mother ends up rejecting her own lamb. Especially since new-born lambs are just about the cutest things imaginable, that outcome is really painful to watch. I had one ewe who lacked some component of this ability. And I say "some component" because a brief phrase like "lamb identification pro-gram" actually encloses many elements, some operating within the brain and oth-ers in the peripheral nervous system. Identification of the newborn, for instance, depends on the ewe's detection of her lamb's odor while licking her newborn off. So if the mother cannot *smell* her lamb in the minutes after birth, or if she cannot imprint her offspring's unique olfactory profile in her brain's *memory* apparatus, then maternal bonding will not occur.

Once bonding does occur, however, the mother has to make many choices during the first year of development. How far should she let her lamb wander, for instance, before calling it back with a distinctive *bah?* As one might intuitively expect, she loosens the reins as the lamb grows, although she calls the lamb close again if something different is afoot (*e.g.* a visitor coming around). Our present point then is that the 'maternal care' program of sheep is flexible. In that same vein, older mothers often provide a higher quality of maternal care, suggesting strongly that they've benefited from their earlier experiences.

Perhaps the *weighting* that's involved in the mother's choices will come out more clearly if we look at the need to select between incompatible programs. For starters, sheep have a 'feeding program'. They will graze for hours at a time on meadow grass (and not just because they're anxious to help me keep it cut). They also have a stay-with-the-flock program that's extremely strong—to the point where they become agitated if separated from the others and will exert themselves maximally to get back with the group. These two programs usually operate happily together. But consider the situation that arises when one of the lambs wanders too far and gets confused. It will let out its distinctive *bah* of distress, and now the mother has a choice. Should she stay with the flock and continue to graze? Or should she stop eating, leave the group, and fetch her lamb? As one might expect, the 'maternal care' program will almost always trump the others in this situation, but only when the weighting mechanisms are properly in place. And clearly, there'd be no sheep to span the next generation if this weighting apparatus were non existent.

MOTHER/CHILD INTERACTION

When a ewe bahs commandingly toward her lamb, the mother displays behavioral expectation that its offspring will respond. And whenever a lamb lets out its own bah of distress, it assumes a similar position of expectation. Not surprisingly to us, the anticipated behaviors are usually forthcoming. Humanly, we might say that the ewe expects her lamb to come when called, and the lamb has equivalent expectations of its mother (something we might express in words like "The mother *should* come when its lamb has gotten into difficulty, because it's her child.") In actuality of course we don't know what if anything the two of them are thinking about in this situation, certainly nothing that's enveloped in words. But we can at least say that, behaviorally, the *bah*s evoke a mother and child reunion that's only a moment away (thanks Paul for that line...and for all your songs).

Because maternal care programs are so central to the existence of mammalian life, and because they so readily evoke in us the notion of *'should'*, my guess is that the maternal care program may represent one of the key forerunners to our human *sense-of-fairness,* which forms the very core of our human ethical program that we will now begin to examine.

14

THE HUMAN SENSE-OF-FAIRNESS

As a young man at Harvard Medical School during the middle of the last century (whew!), I came across the skull of Phineas Gage for the first time. He was a young man who'd had the misfortune—during the middle of the century before *that*—of having a crowbar explode right up through the bottom of his skull and out through the top on the other side. Despite considerable damage to his frontal brain tissue, and in an era before antibiotics, Phineas managed to go on living for a number of years. Few would consider his miraculous escape from death a stroke of good fortune, however, because he emerged from the accident with a Jekyl-to-Hyde personality transformation. Before the explosion, he'd been a likable guy and a dependable worker. Afterwards (if I may be allowed to characterize his behavior as antiseptically as possible) although still mentally alert, he gave undisciplined vent to his baser instincts and alienated all those who tried to befriend him.

His doctor, John Harlow, studied the case carefully and presented his findings to the Massachusetts Medical Society—an excellent instance of research by a busy practicing physician—but Harlow's conclusion that frontal brain injury had destroyed the proper balance between Phineas' intellectual faculties and his "animal propensities" was virtually ignored at the time. Why? Although mid nineteenth century physicians knew that the brain underwrote basic functions like sensation, movement, and even speech, they were still beholden to the then standard philosophical notion that our higher faculties were 'spiritual' in nature and could not emanate from a merely physical object like the brain. (We touched on this issue earlier, in Chapter 6, and will do so again in greater detail when we get to PART III.)

In the century and a half since that time, however, sufficient data has accumulated to show that even the most 'spiritual' of experiences (*e.g.* even the lofty experiences of mystics) are underwritten by specific brain activities.[1] Indeed, without a functioning brain, there's <u>no</u> experience whatsoever and <u>no</u> behavior at all. Restated for emphasis, the tight fit between brain activity and behavior—even behavior that falls into the elevated realm of morals—has by now become abundantly clear. But Doctor Harlow had been ahead of his time. Or perhaps more accurately stated, he was an early harbinger of the era that was just beginning to emerge.

ANOTHER STORY FROM ANCIENT GREECE

Before the notion of Natural Selection came along, the leading explanation to account for the many adaptive features of organisms (up to and including our human "voice of conscience") involved hands-on design attributed to humanesque gods. My favorite story of this genre was told by *Protagoras*, an ancient philosopher who jousted with Socrates in Plato's dialogue of the same name. Here's a summary:

Once upon a time, the gods decided to amuse themselves by fashioning mortal creatures from earth and fire. However, since the gods themselves preferred to occupy their time with play, they assigned the task of drudging out the details to *Prometheus* and his sidekick *Epimetheus*. It was the latter who actually equipped each kind of creature with suitable powers. For instance, some he armed with weapons like strong muscles and sharp teeth, while others he provided with speed so they could avoid danger. To some of those who were small, he provided underground dwellings for their safety, and for others of them, wings for flight. To those whose stature he amplified, their very size formed a protection. He also contrived additional elements like fur as a defense from the cold, and he provided for different sorts of foods. Some creatures grazed on grass, while others ate fruit high off the trees, and some he allowed sustenance by devouring other creatures—though he was careful that these last were made less prolific than their more fertile prey. By such suitable devices he made provision that no species would be destroyed.

Alas, by the time he came to the human race, *Epimetheus* had used up all his ingenuity, so man was left uncovered, unshod, unbedded, and unarmed. Fortunately, however, *Prometheus* stepped in at this point and filled some of the gap by providing our species with the ability to speak and name things and thence to figure out how to clothe and house itself and how to get food from the earth.

But there remained one final and apparently insurmountable problem: The individual human proved no match for large and ferocious beasts of prey; yet when men sought to live together in protected communities, they injured each other so severely that the total destruction of our race seemed close at hand. So Zeus finally had to personally intervene, ordering Hermes to infuse men with a sense-of-fairness, thus enabling the human race to live in sufficient harmony that it has managed to survive till this very day.

DESIGN BY NATURAL SELECTION

In this ancient story, the honcho of the gods (Zeus) delegates most of the work, and his assistants figure things out as they go along, much as we humans do. In our present age, all but Fundamentalist believers also accept the notion that God uses "assistants" to get the job done, albeit not the humanoid sort depicted by the likes of *Epimetheus* and *Prometheus*. However, if we were to personify the modern day helpers in order to enhance our own poetic sense, we might name them *Change* and *Selection*.

In the course of an organism's recurrent generations, *Change* performs his assigned task by seeing to it that genetic copy-errors occur from time to time. Each copy error in turn leads to variations around the theme of a given organism, and the variations that turn out to be most adaptive to that organism's current world will have a leg up when it comes to survival and reproduction. Consequently, their progeny will become more common…*uh*…thanks to the work of our second personage, *Selection*.

And while having no wish to knock *Epimetheus* or *Prometheus*, I must say that these two new characters, *Change* and *Selection*, win hands down when it comes to cleverness. For instance, when called upon to achieve a balance between predator and prey, they never even bothered to invent an *ad hoc* device so that predators would be "less prolific than their more fertile prey." That's because *Change* and *Selection* hit upon the fact that predators starve to death if they eat too many of their prey. So *Change* and *Selection* achieved balance just by letting the number of prey dictate how many predators could in fact make a living over their dead bodies.

Furthermore, *Change* and *Selection* have also been astute enough that their God—unlike the Zeus of old—has never had to micro manage the situation by involving himself hands-on in order to "infuse men with a sense-of-fairness." That's because *Change* was astute enough to introduce continuing variations into genetically encoded behavioral programs, and *Selection* did the rest.

NO *SENSE-OF-FAIRNESS*, NO HUMAN RACE

Some years ago, as part of my effort to continue my education in ongoing fashion, I pushed myself—with a zeal I can only describe as masochistic—into reading some of the stories written by the Marquis de Sade, a collection of his personal fantasies that had been passed along for generations by my enlightened culture. The Marquis, I should perhaps add, never wrote a saga. And with good reason. His collection of psychopaths, seeking only the pleasures of lust and power-over-others, had absolutely no interest in treating their fellows fairly. Living communally with his scoundrels would've made life in Margaret Mitchell's Atlanta immediately after the Confederate Army abandoned the city seem tame and orderly.

More centrally even, since those among the Marquis' characters who became parents would've had no feelings of affection or sense of obligation toward their young children, no offspring would have survived to participate in sagatorial sequels. Bottom line: Amazing as it is that our wondrous species came to be, it's no mystery at all—once given the fact of our existence—that we have a sense-of-fairness program built into our nervous systems. Without such a program, no highly sophisticated social species like ours could ever have gotten off the ground.

THE *SENSE-OF-FAIRNESS*, A FAILED PROGRAM?

Before getting to the specific ways in which our sense-of-fairness program operates, it may be helpful to take a look at one apparently critical problem. Namely, our human sense-of-fairness often appears so feeble and malnourished that it hardly seems up to performing its job. I suspect, for instance, that Bill Gates might not have wanted to begift such a program with the Microsoft name. So how come its manufacturer didn't go out of business long ago? Or alternately stated, how come *Homo Sapiens* is still hanging around, considering that on balance we seem to treat our conspecifics much more unfairly than fairly?

The first point to note, when addressing any instance of evolution, is that Natural Selection has never promised us a rose garden. Or to put it less metaphorically, random genetic variation doesn't necessarily provide the optimal solution to *any* problem. In order for a particular variation to be selected, it has only to function well enough **1.** that the organism so equipped will have a realistic potential for survival, and **2.** that the particular change will prove more effective on a relative scale than the other variations that happen to become available during the same period of adaptive competition.

A "FAILED PROGRAM" THAT GETS THE JOB DONE

Let's illustrate that point with yet another example from our visual system. Recall from Chapter 3 that scientists often use explanatory models patterned after devices of our own human design. In that context, we noted that "Just as the camera has a lens in front to focus an image on the film placed along the camera's back wall, so the human eye has a lens that focuses light on the retina attached to *its* back wall." We saw also that the retina processes light energy in a far more sophisticated fashion than does film emulsion. That's because information-processing 'chips' within the retina (horizontal cells, bipolar cells, amacrine cells, ganglion cells) perform immediate computations over the detected data before the complex 'wiring' (extensive webs of nerve fibers) eventually funnels results to the brain.

Now, one does not need an advanced degree in engineering to figure out the best arrangement for all this micro equipment. Light waves should obviously reach the battery of detectors (the rods and cones) with as little interference as possible. So the rods and cones should be placed at the retinal *surface* where light coming through the lens of the eye will impinge directly on them. The complex 'wiring' and associated 'microchips' should then be set *behind* these detectors and nearest to the main trunk of wires, the optic nerve, that will transport visual information back to the brain. If you were stupid enough to reverse this order by putting the detectors *behind* all this other stuff, the wires and chips would obviously interfere with the free flow of light onto the detectors.

When I read about the actual setup, however, my textbook made it seem as if this backwards arrangement was the actual one. Naturally, I told myself *that* couldn't be the case, and I ended up cussing the textbook for stating the facts so unclearly that the situation got all confused (a credible response perhaps, given the fact that it's unfortunately not rare for textbooks to get a bit muddled). However, on this occasion, my textbook turned out to be exactly on target. The confusion arose from my difficulty in believing that the retinal design simply *is* sub optimal!

Fortunately—and similar to what often happens when humanly engineered devices end up with design bugs—structural accommodations have been set in place that compensate in part for the initial flaw. Turns out that our clear vision comes from only one tiny sweet spot on the retina, the fovea, and in this region the clutter of 'wires' and 'chips' have been swept diagonally so they're mostly out of the way. (Our ordinary sense of having a wide visual field that's perfectly clear comes from the fact that our eyes make frequent darting movements, called saccades, which cause the fovea to repetitively scan the visual field.)

But at any rate, this early exposure to sub optimal biological design cured me of my naive sense that the human body's mechanisms are inherently optimal. Turns out of course that they just have to be good enough to get the job done—and our visual system eventually succeeds pretty darn well at its task.

A MATTER OF *PROGRAM-BALANCE*

Though our sense-of-fairness program may well then be sub optimal, it's not quite <u>as</u> defective as it sometimes seems. In order to start making that point, let me go back to the case of a ewe rejecting its new lamb. I must admit to feeling initially outraged at such abandonment—a mother darn well *should* attend to her baby's needs—but when I finally manage to get a grip on myself, I'm able to realize that once the recognition and bonding programs fail to kick in, the mother subsequently behaves appropriately. After all, given her limited supply of milk, she 'should' save it for her own identified child, and not squander it on this stranger.

To emphasize that point, consider the following situation in our own species: Suppose a mother were to allocate her available resources of time, money, and emotional energy equally among all the children in her town. We might well end up taking her to task for neglecting her own children. For if she used her limited grocery money to give food to *all* the children, there might not be enough left for her own. And if she spent her time baby sitting *all* the children in the neighborhood equally, she would likely not have enough time to adequately supervise her own.

That would occur rarely, in fact, because of a built-in 'rheostat' that energizes the sense-of-fairness program variably. To explain how that comes about, let's first express the gist of the program in words: *"I should be treated fairly, and I should treat others who are <u>like me</u> fairly."* The *"like me"* amounts to a crucial qualification, since it leads to gradations in weighting. The more others are *"like me,"* the more strongly the program kicks in—and not surprisingly, one's own children usually kick the program into high gear. In the coming chapter, we will examine this crucial weighting element in more detail. But for the present, I want to underline a more limited point: If the 'fairness program' energized responses to every applicable situation with the same degree of weighting, the program would become seriously maladaptive. Or to apply an Aristotelian dictum: Any virtue becomes a vice by *excess* as well as by deficiency.

15

"LIKE ME"

Our crucial behavioral programs are not coded in language, though of course when we talk about them, we do so by trying to represent the gist of a given program in words. That's what I was about in Chapter 14 when I rendered our sense-of-fairness program by the statement: *"I should be treated fairly, and I should treat others who are <u>like me</u> fairly."*

Note the asymmetry involved in this statement. People tend to feel strongly that they should be treated fairly by <u>others in general</u>, but they seem to feel called upon to treat others fairly only in so far as the others are *"<u>like me</u>."* And it's in fact the *"like me"* part that constitutes one of the most crucial differences among various languages-of-fairness. For *"like me"* may be pretty much restricted to my own band of gypsies, or to folks of my own skin color, or to those who wave the same national banner, or who belong to the same institutional religion…and so on. Or the sense of *"like me"* may extend to all members of our Global Village.

THE GOOD SAMARITAN

Reminds me of the story Jesus of Nazareth told about the Good Samaritan. Jesus had just gotten through reminding his listeners that they should love God, and that they should love their neighbor as themselves. No doubt wanting to pin the notion down more clearly, one of his listeners asked in effect: "Yeh, but I need to know who exactly is to count as my *neighbor*." So Jesus proceeded to tell a tale about this fellow who got robbed and beaten to a pulp by a bunch of thugs. Some folks from the fellow's own ethnic group, including an official who certainly should've acted more responsibly, just continued about their business, leaving him all banged up in a ditch. But then along came this other fellow—not only from a different group, but from a group that was looked down on—and he felt pained about the victim's plight. So the guy (from this minority group mind you) helped the beat-up guy big time, and Jesus ended up asking his audience "So who was *neighbor* to the fellow who got beaten up?" Every time I hear this story, I figure Jesus was trying to push his listeners to develop a more encompassing sense of who should be placed in the category of *"like me."*

Of course, if you go to that same Biblical treasure-trove, you can find a bunch of other stories enabling you to narrow down those who rate a *"like me"*—stories, for instance, in which the *"like me"* guys trick the outsiders by their cleverness, and other stories in which the outsiders betray the trust of the good guys in order to gain perfidious advantage. Same bag of tricks each time, but different ways of describing the situation, depending on whose ox is being gored.

WEIGHTINGS AND COUNTER WEIGHTINGS

I suppose, by the way, that I've just been caught with my preachiness hanging out, but that's an occupational hazard for all those who investigate the arena of human values. Before we even realize it, we find ourselves slip-sliding from the notion of values-in-general to our very own ideal values. And to make things even more difficult, the *ideal values* that we espouse with great verbal ardor are not at all necessarily the weightings that show themselves most heavily in our actual behavior. That's because differently weighted programs often compete for expression. Recall, for instance, that the ewe's maternal care program will at times have to weigh in against feeding and flocking programs.

Or take the instance with which I opened Chapter 13. Since I share Albert Schweitzer's "reverence for life," and since it's only fair to respect a life that one reveres, how come I whap at deerflies when they land on me? Answer: I also have a highly weighted self-protection program that's competing for expression—not to mention that I respond to the deerfly as so little *"like me"* that my sense-of-fairness program hardly gets off the ground.

ANIMAL RIGHTISTS

Some folks do, however, find their *"like me"* response extending strongly toward animals, or at least to the mammals and birds that are voyaging with us on our modern day Ark. And truth to tell, these Animal Rightists have influenced my own language-of-fairness. For instance, I no longer order Veal Marsala, which was one of my favorite dishes, because I agree that it's unfair to a young calf to allow it to be born, and then pen it into an almost motionless state in a dark barn until its slaughter—one thing to eat meat, another thing to partake of a young grazing animal that's been so cruelly fettered during its entire life.

Animal Rightists, by logical extension, tend toward vegetarianism. However, if their language-of-fairness extended even more widely and even more strongly—enough that it encompassed a radical reverence for *all* life—would they not also refrain from eating vegetation? But in that case, they'd starve to death. So it's a good thing for our species that the fairness program, at some point, does indeed taper (just sad that it so often tapers so markedly, so soon).

Now here's a personal instance extreme enough that I'm a bit embarrassed to mention it. When I cut down a tree, which I do frequently enough since I heat with wood, I have to overcome qualms that arise unbidden. That's because my spontaneous *"like me"* response seems to extend a tad to these beautiful instantiations of life—more, for sure, than to animals of the deerfly, tick, or mosquito ilk. But I'm confessing this peculiarity of my *"like me"* response, not to prove I'm some kind of weirdo (let me reassure readers that I practice normality for at least 30 minutes every day), but to emphasize once again that fairness programs taken to extreme lengths would become seriously maladaptive. (And by the way, since almost no one feels that a need for fairness extends beyond *sentient* life, let me note that we will address in **Part III** the issue of where phenomenal consciousness leaves off.)

ROPE DANCING THROUGH LIFE

To summarize then, it's highly likely that the sense-of-fairness program, as historically selected on our planet, *is* less than optimal. It's also the case, however, that if the program did not have a built-in rheostat that gradually dimmed this sense, then our species would not have been able to survive. And even when dealing with our fellow humans, the rheostat requirement is essential. To state the resulting paradox explicitly, a normally operating sense-of-fairness program delivers the following dictum: *It would be unfair to be equally fair to all.* Various languages-of-fairness will nuance this dictum differently, but no successfully adaptive program lacks this imperative. Hence, most people whom we'd think of as "fair minded" are required to perform a continual balancing act through life.

THE FAIRNESS PROGRAM'S CHIEF COMPETITOR

For most of us, there's little danger that our sense-of-fairness will run roughshod. That's because our 'self-interest program' has more muscles than the youthful Arnold Schwarzenegger. Its components go far beyond the critical drive for self-preservation to include items like personal comfort, pleasure, and self-esteem. (The last, by the way, is by no means just a luxury; if a person's self-esteem falls literally to zero, he will not take the ongoing measures necessary to sustain his own life.)

Components of the self-interest program, acting in concert, seem easily capable of overwhelming the average person's sense-of-fairness. That's perhaps why our herd sense, naturally selected as an additional counterweight, tends to strengthen our individual sense-of-fairness by providing praise for adherence and reproach for transgressions (a point we will be focusing on in the coming chapter). Even more, we are programmed as a species to *admire* behavior that goes beyond fairness to altruism. We tend naturally to venerate heroes who sacrifice themselves for the greater good, especially when we more ordinary folk can…*uh*…do so at an unsacrificing distance.

THE CONVOLUTED LEARNING OF 'FAIRNESS'

In the Introduction to PART II, we noted that the particular language-of-fairness learned, as in the case of ordinary language, will depend on that displayed by one's familiars. Just as those growing up among English speakers will learn English, those growing up within a Christian culture will tend to learn a Christian language-of-fairness. But the analogy starts to limp for the following reason: We can almost always count on it that if a child's close familiars speak a given language, the child will grow up speaking that language first and foremost. By contrast, children have been known to end up with a primary language-of-fairness quite distinct from that of their parents.

Franklin Roosevelt and Jack Kennedy could be used to illustrate the point. Both grew up in English speaking families, and both ended up as English speakers (*"Ich bin ein Berliner"* not withstanding). But both were also born to affluence—and with it the elitist version of fairness that develops from natural self interest—yet both ended up with a highly egalitarian sense of distributive justice. That's why earlier we had to append the important qualification that the language-of-fairness seems to be learned in more convoluted fashion.

The reason for a more unpredictable result isn't totally clear, though pursuing the analogy with language may still shed some light. Take the case, for instance, of an immigrant child whose immediate family speaks a foreign language. The child will of course learn this language, but he will also learn the language used by his new peer group and the rest of his extra familial relationships. Intuitively, he may then start to favor his new language until it becomes his primary mode of communication—one that's fully congruent with his growing sense of identity as a citizen of his new land.

Something like that can happen to one's language-of-fairness as a child's continual growth brings him progressively beyond the confines of his family and immediate subculture. The opportunity to connect with a wider world often evokes the pleasant sensation of belonging to a "larger family," and the sense-of-family can in turn evoke an actually comforting sense of "like me"—especially when there are no immediate disadvantages. In that respect, people of privilege are sometimes able to make such a change more readily, because they don't experience direct downsides like having to compete with "the others" for laboring positions. On the other hand, in so far as the privileged are careful to stay within their own birthright bastions (right school, right club, right work place…uh…if any), they remain protected from fully extending their sense of "like me" to their less well-funded fellows.

ANOTHER COMPARISON WITH LANGUAGE

And there's an additional similarity to language, in that both change gradually over the generations, sometimes in response to significant social circumstances. Us macho guys, for instance, are careful nowadays to say things like "He hit'm real good" when a line backer makes an impact tackle. Only an effete intellectual (or worse) would resort to the proper adverbial form in such cases. Or resort to wimp phrasing like "We macho guys" for that matter. And though it's exhausting to have to practice our swaggerizing, both physically and verbally, day in and day out, I'm sure the rest of you *real* guys out there will agree with me that it's well worth our effort.

Of course, there are those who seem to make such verbal adjustments effortlessly. That's something I've always admired about national news anchors like Peter Jennings. He drops a bunch of his adverbial endings with an ease that'd make you swear that an *'ly'* suffix never even existed. Or maybe he's hit upon the fact that we have to simplify our language somewhere in order to compensate for the ever enlarging vocabulary we have to lug around (so we can label our fancy new consumer products, not to mention the ongoing deluge of scientific discoveries). And after all, since the word that an adverb modifies usually *is* evident from the context, this increasingly common speech-simplification will probably reign supreme before too long.

In like manner, the language-of-fairness changes over time. It seems almost difficult to imagine now that less than a hundred and fifty years ago, the citizens of our Land were seriously split on the issue of slavery, with many of us convinced at the time that there was no inherent unfairness in its practice. Now, virtually all of us would consider slavery totally repugnant...unless of course its practice be confined to sweat shops in third world countries.

HOW CHANGES OCCUR IN THE "LIKE ME" RESPONSE

The above instances involve gradual change in population-wide expressions of highly complex behavioral programs, but note that the alterations require no change in genetic coding. In fact, it's this similarity that allowed us to make ready analogy between ordinary language and what I referred to as the language-of-fairness. Let's take a moment now to review the distinction made in Chapter 13 between *preset* and *programmable* programs. We noted that the former come already wired, so they operate rigidly (like our simple hand-held calculators). Highly complex computers, however, can be additionally programmed to learn from experience.

Preset programs have one terrific advantage: They're pretty much up and ready from the start, and their outcome is automatic. We might use the stretch reflex as an example. Reason the knee jerk's so predictable is that it involves a simple built-in arc between an incoming nerve and an outgoing nerve. (We might further note that we often extend the phrase "knee jerk response" metaphorically to conceptual responses that occur quite predictably in persons of vehement convictions.)

But this very predictability of built-in responses limits their adaptability. To illustrate: Imagine our language composed of prewired symbols like the three identified alarm calls of the Vervet monkey (Chapter 5), except we'll picture ourselves gifted with 50,000 such symbols. We'd inherit then an extensive (not to mention effortlessly obtained) vocabulary, gradually accumulated over probably hundreds of millennia. Only problem is, we'd have no way of referring directly to modern stuff like TV sets, thermostats, automobiles, hydrogen atoms, space capsules, blue jeans, lawn mowers, and on and on—or at least we'd have to wait on chance DNA changes for additions to our word list.

But coming to our rescue like the Lone Ranger, the exceedingly complex human brain has been able to support the operation of programs that become fully up and running only when provided with extensive environmental input. This permits relatively rapid change, even within a single generation, and in such a way that whole populations can soon be provided with the same programmatic innovations.

A CONSOLATION FOR BELEAGUERED PARENTS

Two opposing factors inherent in our programmable programs are important to mention at this point: the tendency toward *conservation*, and equally crucial, the tendency toward *change*. Parents readily discern both tendencies as their children grow, though the latter tendency can be a pain in the butt when it results in unpleasant rebelliousness. Sometimes changes are acceptable enough, as when youngsters coin hip new words that the older generation has to catch up with ('hip' was a hip new word in my day), or put rings in their noses and tongues in order to be properly stylish. At other times, the changes are more troublesome, as when they smoke marijuana, instead of letting White Lightning remain the biggest thrill of all. Old fogy style.

Consider the outcome for our species, however, if *conservation* ever did reign supreme. Children would be more happily biddable of course, but *change* would come mighty slow (I mean slowly), so we'd probably still be living in the equivalent of some ancient era. And the fact is, we might not even have been behaviorally agile enough as a species to survive the huge and relatively acute environmental challenges with which our species has already had to cope—like the Laurentide Glacier that a little over 10,000 years ago still had a thousand yards of ice dumped on top of the patch of land where I'm currently sitting. And we've certainly come a long way health-wise too, when you consider that the majority of us now live years beyond the three-score-and-ten that was considered the natural human lifespan in days of yore (when *actually* making it to age 70 was quite unusual).

SUMMARY

Our sense-of-fairness then emerges from the operation of a highly complex programmable program; hence, it requires lots of environmental feedback before becoming fully operational. Crucial input comes from family and others within one's group—albeit the group is capable of enlarging as one's world expands. Most such programmable programs are highly *conservative* on the one hand, yet on the other hand they have a significant tendency toward *change*. The latter tendency works more actively, all else being equal, when the nervous system is relatively young and pliable, accounting for both the rebelliousness and the

<u>innovativeness</u> of youth. In contrast, the trend toward conservatism and stability (some would say ossification) tends to become more pronounced as we age.

In the upcoming chapter, we will now begin to address the ways in which our sense-of-fairness asserts itself at the level of felt motivation.

16

WHENCE THE MOTIVATIONAL POWER OF OUR SENSE-OF-FAIRNESS

As a prelude to explaining how our sense-of-fairness program manifests itself at the *motivational level,* let's review a couple of background notions. Recall first from Chapter 2 that scientists aren't much into chasing the Holy Grail of absolute truth. Instead, they gather useful data with an eye to formulating conceptual models of reality, approximations that can provide a reasonable explanation for systematic observations and a reliable basis for predicting future results.

The more broadly a particular theory applies, the more that theory is valued. And when a successful theory can be formulated in relatively simple and strikingly bold strokes, scientists are wont to apply an esthetic judgment by referring to the theory as "elegant." That's because when scientists contemplate such a theory, they tend to experience the sort of subtle pleasure we associate with, say, viewing a beautiful painting or reading a favored poem. Esthetics aside, however, if a theory were too complexly detailed, it would be of little use to us despite its accuracy. Every helpful conceptual model must be suited to our human limitations.

With that perspective in mind, the theory we're about to make use of—Skinner's theory of *Operant Conditioning*—is not only useful but also elegant. That's because it applies very broadly within its field of application and leads to many successful predictions, all the time dealing with reality at a level of complexity that far exceeds the basic level investigated by particle physicists. Furthermore, the theory is simple enough to be briefly expressed in a few bold verbal strokes: *Behavior that's positively reinforced will occur more commonly; while behavior that's negatively reinforced will occur less commonly.*

In order for this Theory to apply, animals require little more than a nervous system capable of forming durable links between behavioral output and what follows. Not that the theory is capable of providing a satisfying explanation for *all* aspects of behavior. Recall, for instance, from Chapter 5 that Skinner's attempt to explain language acquisition on the basis of Operant Conditioning was not fully up to its task (even though associative mechanisms *do* receive important play in the linguistic arena, especially when it comes to vocabulary formation). For our present purposes, however, a modified version of the theory will prove helpful in explaining how our sense-of-fairness program does its work.

SKINNER'S THEORY

One of the attractive features of *Operant Conditioning* is the fact that it codifies common sense: Other things being equal, reward an activity, and it *will* be more likely to happen again; punish an activity, and it *will* be less likely to happen again. But in order to understand the strictures Skinner placed on his definition of the term 'reinforcement', it's useful to review the historical context in which he formulated his theory:

In the first part of the Twentieth Century, the disciplines collectively referred to as Behaviorism became progressively favored by experimental psychologists. That's in large measure because the preceding era of psychology had gotten bogged down in endless accounts of subjective experiences, and such "introspectionism" led to results that were hard to reproduce. By contrast, external conditions of an experiment were highly controllable, and externally directed observations were relatively straightforward. So Behaviorists limited themselves to examining behaviors as they occurred in settings where the external variables could be individually manipulated. In Skinner-type studies, for instance, a rat would occasionally trip a lever as it wandered around its laboratory cage. This "emitted" behavior could then be sequenced with another event of the experimenter's choosing (e.g. a food 'reward'), and the subsequent frequency of bar pressing could be accurately measured, leading to results that were both predictable and quantifiable.

And there was an additional factor of great practical importance at the time. So little was known about how the brain worked that it made sense to treat the organ like an unopenable "black box." How this enigmatic item gave rise to behavior was a mystery, but external happenings did in fact influence the occurrence of the particular behaviors that the black box emitted. Furthermore, these external happenings could be closely controlled during experiments, and the subsequently emitted behaviors could then be quantified.

SKINNER'S THEORY MODIFIED

Notice, however, that external reinforcers in no case work directly. They produce their actual impact only when registered by the animal and processed by that animal's internal weighting apparatus. Food pellets, for instance, work as positive reinforcement because of a rat's appetitive system; electric shocks to its feet work in the opposite direction because of the rat's nociceptive system. The effectiveness of *external* reinforcement, that is, depends on triggering *internal* reinforcing systems. Furthermore, that fact always entered into the equation of experimental reproducibility. For instance, rats who'd already eaten weren't about to respond with quite as much alacrity when a particular behavior was reinforced by food, because the hunger drive's weighting varies in accordance with internal variables like degree of satiety.

And as far as the *internal* reinforcing systems are concerned, progress continued to be made during the second half of the Twentieth Century in identifying actual motivational circuits. For instance, when neuroscientists learned how to provide reinforcement for lever-pressing behavior by arranging *direct internal stimulation* (via electrodes implanted in the septal region of the rat's brain), these animals increased their lever-pressing activity to proportions previously unheard of with the use of merely external reinforcements like food. Bottom line: Since all external reinforcement works by evoking the activity of internal reinforcement systems, and further, since the brain can no longer be adequately thought of as merely a "black box," it makes sense to call upon *internal* as well as *external* levels of reinforcement in order to explain how our behavioral programs work. And that's what we will do now as we're providing a motivational explanation for how our sense-of-fairness program gets its job done.[1]

A POTENT PAIR OF INTERNAL REINFORCERS

Let's illustrate what happens by following a Country-song hero as he makes his way righteously through the lyrics. The song I have in mind starts with our married hero drinking in a somewhat disreputable tavern. We know it has to be disreputable because a barstool beauty's busy turning her blue eyes into green lights in his direction. Naturally, his will to resist is weakening pretty fast, so he owns up to the fact that "on the one *hand*, I could stay and be your lovin' *man*." But then he realizes that "on the other hand" [which saying he interprets very concretely], he has a "golden *band* which reminds me of someone who would not under*stand*." Happily, virtue triumphs, and he decides to head on home.

By the way, I've italicized the rhymes above so readers can enjoy the acoustic pleasure flowing from the use of that device. As for purists who might claim that *"man"* as it occurs in the above lyrical flow amounts to no more than an imperfect rhyme, let me remind you that no decent Country singer would be caught dead hammering home the 'd' sound at the end of words like *"hand"* and *"band."*

But back to the point, our hero's language-of-fairness tells him that it'd be *unfair* of him to cheat on his wife, and his sense-of-fairness makes its case in a highly characteristic way: He feels guilty as hell. Hence, pangs of guilt help outweigh the anticipatory pleasure he's experiencing. Alternately stated, his guilt feelings act as a potent negative reinforcer[2] in the face of his contemplated behavior. And there's a subtle yet powerful *positive reinforcer* also at work. For when he successfully resists the temptation, he also experiences a feeling of pleasure at having done the right thing. For reasons we'll assess later, we don't seem to have a ready label for this genre of positive feeling, so let me scoop up a phrase from an old nursery rhyme to make the point. I'm referring to Little Jack Horner when he ended up with the refrain *'What a good boy am I'."*

Note that these two reinforcers act as carrot and stick. Our hero experiences positive reinforcement from the pleasure he feels at adhering to the right, and he experiences negative reinforcement from the guilt evoked by his urge to do evil.

A POTENT PAIR OF EXTERNAL REINFORCERS

And there's more. If our hero had indulged himself, the chances may have been small that he'd be discovered. Nevertheless, it was possible that his wife and family might find out, and the mere possibility was sufficient to evoke an imaginative scenario wherein he becomes publicly *shamed.* I refer to shame as an external reinforcer, therefore, since it involves the presence of a disapproving agent external to oneself (if only in contemplation). By contrast, guilt comes unbidden completely from within, even if no outsider knows or can know about the guilt-provoking activity. Overlapping feelings of shame and guilt together comprise a hefty negative reinforcement.

On the positive side, there's an equivalent overlap, because the affirmation others often provide in response to doing the right thing evokes considerable pleasure. Thus, as the character in this song recounts his adventure, he's pleased to receive the implicit accolades of his audience for ultimately adhering to the right. Interestingly, we also don't possess a ready label for this genre of positive feeling, so we'll make up an expression, *pleasure-at-being-praised.* Analogous to the earlier instance of negative reinforcement, the overlapping sense of *What-a-*

good-boy-am-I and *pleasure-at-being-praised* together comprise a hefty positive reinforcement.

In summary, although all reinforcements eventually produce their effect through internal reinforcing mechanisms—and in this sense *all* reinforcement is ultimately internal—some reinforcements involve the occurrence of external events (recall that Skinner made this a requirement of his original theory). In that context, we're referring to the affirmation or condemnation by those who are outside the self as *external* reinforcement. To complicate matters, however, when it comes to an extremely complex organism like *Homo Sapiens*, what I'm referring to as an "external" occurrence may on occasion happen only via the imaginary projection of what *could* happen (another instance of our ability to let the products of our imagination die in our stead). Note by contrast, however, that the reinforcement of guilt can occur without *any* reference to external condemnation, as instanced by the case of a person feeling guilty when he entertains forbidden fantasies of a purely private nature. That's the reason for characterizing guilt feelings, in the present context, as *internal* reinforcement.

A PROBLEM

There's a problem, however, with the sort of explanation I've just provided; namely, many *many* upright folk judge that they try to do the right thing just because it *is* the right thing to do. Period. And in fact it seems to them more than a bit demeaning that one should be pushed and pulled into doing the right thing by merely animal responses like pain and pleasure—the latter stance being referred to by ethical theorists as 'Hedonism'. Since the above explanatory story does indeed fall into this oft rejected category, let's examine next what's actually involved in hedonist theories.

17

HEDONISM

Why is it that doing something simply because it's the right thing to do, or avoiding something simply because it's the wrong thing to do, seems more noble than achieving a similar outcome via the intercession of pain/pleasure mechanisms? Perhaps because operating in a fashion that enables us to obtain pleasure or to avoid pain strikes us as reasonable, but not especially *admirable*. By contrast, a person who performs good deeds from purely spiritual motives seems to be operating on a far higher plane. At least, that's what many of my informants say.

But if we ask what makes a motive "spiritual," answers tend to become rather nebulous. When pressed, here are some sample responses: Those who frame the basic issues of life within the theology of our institutional religions tend to say things like "An example of a spiritual motive is when you do something in order to follow God's law." If pushed to explain why anyone is actually *motivated* to follow God's law, a practical response will sometimes be forthcoming: "If you obey God's law, you go to heaven; if you disobey God's law, you go to hell." Clearly, this explanation does no more than push the notion of reinforcement to the hilt—a colossal carrot and stick constructed of eternal joy and eternal pain.

Scholastic theologians, being quite sensitive on the point, made a distinction between the above motivations (worthwhile though hardly optimal) and the far loftier motivation provided by a pure love of God. But consider: Does it not give us pleasure to please those whom we love…and the greatest pleasure to please one whom we love most of all? So we have not avoided the notion of reinforcement, even in the case of this high-ground scenario. And note while we're at it that the dimension of pain is often omitted from this 'purified' explanation. Why? Perhaps because we tend to judge actions motivated by negative reinforcement as not quite so worthy. (Biblically inspired theologians might put it this way: "Fear of God is the beginning of wisdom, but love of God is its crowning achievement.")

NON HEDONISTIC HUMANISTS

Many non theistic humanists also join the ranks of those lacking enthusiasm for the notion that our moral decisions are regulated by pleasure/pain mechanisms hooked up to each individual's sense-of-fairness. In fact, I've been surprised to see how an anti hedonist stance sometimes plays out in the thinking of even highly sophisticated ethical theorists. One well-known secular scholar, for instance, described hedonist self-interest as *"never self-denying."*

Reminds me of a story from the ongoing annals of internal medicine. Seems there was this young man who'd given himself over to the never-self-denying pleasures of a dissolute life. As a result, his health began to fail badly, causing him finally to seek medical assistance. However, the kindly old physician whom he consulted could but shake his head in dismay after completing his examination. "Son," he said, "unless you're willing to make drastic changes in life style, you're not going to be around much longer." The doctor then meticulously reviewed the many alterations needed in regard to diet, alcohol, drugs, sex, and so on—all of which combined to leave his patient blanched by the mere contemplation thereof. Nonetheless, after reflecting on the gravity of his personal situation, the young man finally asked: "Doc, if I make all these changes, can you promise me that I'll live longer?" Not wishing to discourage his patient, though wishing to remain faithful to the truth, the doctor weighed his words a bit before finally responding: "Well, I can't promise definitely that you'll live longer, but I can promise it'll *seem* longer."

First point after you finish chuckling: A reasonable theory of Hedonism would *never* advise one to be *"never self-denying,"* as in the case of this young man, because enhanced pleasure of the moment can often drastically reduce one's pleasure/pain ratio in the longer run—and the goal of Hedonism as a consciously articulated path in life is to optimize one's pleasure/pain ratio.

Second point: Note how even the impulsive young man described shows our normal human tendency—during crunch time—of working in the direction of an improved pleasure/pain ratio (in this case, less fun for much more time in exchange for more fun for far less time).

Third point: Note how the notion of *pleasure* is often, as in the present story, tacitly limited to what Jeremy Bentham referred to more than two hundred years ago as "organical." Contrast this restriction with the obvious fact that some of our most treasured pleasures are highly refined, like our response to viewing a painting by Renoir, our response to hearing the choral movement of Beethoven's Ninth, and, yes, our response to performing a good deed that we're proud of.

Which leads me to my speculation as to why we have simple words like feelings of *'guilt'* and *'shame'* on the negative side, whereas we have no similarly straightforward terms for the positive reinforcement that accompanies our adhering to the right. (Recall that when it came to labeling the latter reinforcements, we had to resort to awkward phrasings like *"What a good boy am I"* and *"pleasure-at-being-praised")*. The lack of clarity engendered by the absence of forthright labels is not likely to be mere coincidence. More probably, it's associated with the fact that so many people are keen on believing that they adhere to the right simply because it *is* the right thing to do. Consider the advantage then of this nomenclatural poverty: When we don't have a handy label with which to highlight an item, that item's more likely to pass beneath the radar of our explicit experience.

ALTRUISM

Along similar lines, most of us have a love affair with the notion of altruism. We judge that when a person does the right thing in order to avoid punishment, it's prudent; when a person performs an act of enlightened self interest, it's good; but when a person like Albert Schweitzer engaged in a pattern of helpful activity that was in no way required of him, at great personal cost, and for no personal benefit, we think of it as totally awesome. Under the circumstances, saying that his charitable work was reinforced by subtle feelings of pleasure might seem to demean his actions in crassly cynical fashion.

Nevertheless, a person's life of steady devotion can hardly be viewed as a series of unmotivated (random) acts. When a saintly man takes quite seriously the Biblical mandate that "What you do for these, the least of my brethren, you do also for me," he is expressing both his love of God and his neighbor as himself. We've already noted, however, that doing something helpful for a person whom we love evokes a subtle yet ofttimes intense pleasure. That's in fact one of the outstandingly adaptive aspects of the 'social program' naturally selected within our extraordinarily complex species: Affiliative bonding behavior comes with built-in positive reinforcement.

What Jeremy Bentham recognized years ago in the social arena was that the genius of noble souls rests in their propensity to experience acutely the sort of subtle positive reinforcement we've just mentioned. Here's how he actually put it in the quaint language folks were wont to use during the seventeen-hundreds: *"A man's moral sensibility may be said to be strong, when the pains and pleasures of the moral sanction show greater in his eyes, in comparison with other pleasures and pains...* *(50)[1]* When it comes to us lesser mortals though, most of the time our more ordinary self-interest programs easily outweigh this genre of reinforcement.

HUMANISTS AND THE 'GOLDEN RULE'

Humanists of all persuasions value the Golden Rule: *"Do unto others as you would have them do unto you."* But, we might ask, what is it that gives this statement motivational power? After all, an alternate rule like *"Do unto others before they do unto you"* may be good for a chuckle, but it fails to activate our sense-of-fairness response. Answer: the Golden Rule is but an alternate way of putting our sense-of-fairness program into words. The statement that we originally used, *"I should be treated fairly, and I should treat others who are <u>like me</u> fairly,"* covers more or less the same ground—differing only in that it emphasizes the primacy of our self-regard, and that it specifies those who are to actually count as "others" within the charmed circle of fairness.

BEHAVIORAL PROGRAMS vs. *STATEMENTS* ABOUT PROGRAMS

Keep in mind, as noted previously in Chapter 15: "Our crucial behavioral programs are not coded in language, though of course when we talk about them, we do so by trying to represent the gist of a given program in words." But whether or not we ever bother to verbally articulate our *sense-of-fairness,* the underlying workings of this program will show up in our behavior and at the level of personal experience (via feelings of guilt and pleasure-at-doing-good). Specifying a program in words *does* help to improve our focus, however, even though statements of themselves generate no behavior; they simply emphasize the directives of an underlying sense-of-fairness that's already up and running. (That's what's involved, for instance, when theorists espouse Hedonism as "an explicitly articulated guide to life").

THE UNIQUENESS OF *HOMO SAPIENS*

Members of our species spend lots of time emphasizing our glorious superiority over the other life forms on Spaceship Earth. Seems to help with the task of shoring up our self-esteem. Go all the way back to our ancient creation stories, in fact, and you'll see our forefathers already busy at work providing themselves with such kudos. The happy account in Genesis, for instance, has God telling the original members of *Homo Sapiens* to live like conquistadors: *"Fill the earth and subdue it, and rule over the fishes of the sea, and the fowls of the air, and all living creatures that move upon the earth."*[2] We get so enthusiastic about this kind of notion that we end up longing, not just for superiority in degree of complexity but also for a qualitative superiority of total uniqueness. And *that* forms another reason we want to resist comparison with "mere animals." It can seem belittling, from such a perspective, to think that our behavior is regulated by the unworthy dross of animal mechanisms like pain and pleasure.

But even within a hedonist model, our species is still pretty darn special, given the remarkable machinery involved in our choice-making behavior—dependent on numerous systems of extraordinary complexity that interact simultaneously across multiple levels.

THE ABILITY TO CONSTRUCT *'VIRTUAL REALITIES'*

Let's illustrate this fact by highlighting just one of our amazing abilities. For emphasis, we'll start as usual with the simpler apparatus available to lesser animals. We saw in the case of a laboratory rat that it would bar-press more frequently when provided positive reinforcement and less frequently when provided with negative reinforcement. Now for a question: What do you think happens when scientists administer both positive *and* negative reinforcement in response to bar pressing? Does the rat subsequently press the bar more or less frequently?

One factor that comes immediately to mind involves the relative strength of reinforcements. We might guess that if positive reinforcement is weak (a mere crumb) and negative reinforcement humungous (a strong shock), then the stronger reinforcement will carry the day. But within rather broad limits, the most critical factor turns out to be the *timing* of a reinforcement more than its ultimate strength. To wit: The reinforcement that comes first will generally win out.

After a bit of reflection, that result won't surprise too many people, least of all the merchants who deal with our fellow species-members on a daily basis. Take the instance of *signing-on-the-dotted-line* behavior. There's an immediate positive reinforcement (we get the item we've been lusting after) and an immediate negative reinforcement (the pain of parting with our money). So "Buy now, with no payments required until next year." The positive reinforcement comes right away, the negative reinforcement only down the line.

Some readers may think it deplorable that fellow humans have been taken advantage of by this blatant manipulation of what scientists refer to as "reinforcement schedules," but my own heart goeth out to the poor merchants. When they try to sell stuff in *cash-on-the-barrel* mode, their upfront behavior is likely to be negatively reinforced by poor sales. But when they change their behavior by offering deferred payment, sales increase. So the altered sales behavior of these hapless merchants gets positively reinforced. (The preceding's a variation on the old story told about laboratory rats who condition scientists to give them food pellets by positively reinforcing this behavior with dependable experimental results.)

In more serious vein though, the crucial importance of timing in regard to reinforcements shows itself clearly in those struggling with an alcohol problem. Friends and family often look on, tragically bemused, wondering why Jim is in the process of throwing away all that he's treasured in life just for a transient series of booze-glows. Problem is, the pleasure from alcohol comes NOW, whereas the loss of family and profession come somewhere hazily down the line.

Now, while these human examples of conditioned behavior show a striking similarity to the behavior of laboratory rats in a Skinner box, differences turn out to be even more marked. We'll focus on one crucial dissimilarity: <u>Our extremely sophisticated representational apparatus allows us to elaborate vividly detailed scenarios that do not yet actually exist</u>. The simpler cerebral cortex of rats has no systems capable of elaborating such vivid *'virtual realities'*, and this factor alone puts us in a class by ourselves when it comes to making choices.

Let's illustrate the nature of this difference by looking at a dramatic instance from history: When Fundamentalists within the institutional religion of the great Galileo took umbrage at his suggestion that earth was not the center of our universe, they needed to deal with a delicate political problem. After all, it's one thing to use the tools of Inquisition on ordinary folk who might contest the...*uh*...truth of our centrality in the universe, but quite another to employ thumb-screw and rack on such an illustrious personage. However, one of those burdened with the solemn responsibility of bringing Galileo back to his senses developed a simple solution to this embarrassing problem. As he put it: "Just show Galileo the instruments." And indeed doing so was sufficient to evoke a *'virtual reality'* within the great scientist, by which he was able to experience the horrific pain and tissue destruction without having to experience the real-world procedures. In marked contrast, a rat would have to be provided with the actual external reinforcement in order to shape its subsequent behavior.

On a more positive note, Alcoholics Anonymous is famous for a format in which speakers detail the horrendous problems that assaulted them as a result of their drinking. One useful result of this intuitive approach was to make *immediately vivid* the sorts of problems that listeners would be facing only at some time in the indefinite future. Alternately stated, the *virtual reality* evoked by graphic tales of horror, related by comrades, provides immediate rather than long-delayed negative reinforcement.

And of course, consumers are *not* totally beholden to the influence of deferred payments, since they have the ability to put "next year's payments" right up front in their imagination; though it's also clear that this ability to experience a *virtual reality* cannot always hold its own with the experience of, say, actually test driving this luxury vehicle right now—especially when other *virtual realities* become factored in, like the status-building impact this item will provide (given the power that symbols exert on fellow humans).

In short, human imagination can provide 'virtual reality' of a vividness far surpassing that available to lesser animals. Combined with our explicit sense of *past/present/future*, this apparatus contributes immensely to our freedom from having to react to a sequence of actual external reinforcements. And since an indefinite series of virtual realities can be elaborated, this ability of itself hugely complexifies our acts of choice, vastly broadening the scope of our pleasure/pain responsiveness—though *not* negating it. Nor would we expect nature to scrap a basic mechanism that had already proven its worth time and time again; especially when one considers that Natural Selection is in its make-up conservative, since new species must originate from variations occurring in already existing forms.

BUT THERE ARE PROBLEMS...

Nevertheless, problems remain with this explanation. And since further analysis will also serve to review some key points that we've touched on earlier, let's next address an ongoing limitation of the conceptual model just presented.

18

LIMITATIONS OF A MOTIVATIONAL-LEVEL MODEL

Seems there was this young doctor in training who felt totally stressed out during his Emergency Ward rotation—so many acutely ill patients requiring critical intervention, yet the limits of his own medical knowledge left him frequently uncertain about how to proceed. He managed to take heart though as he watched a highly experienced physician working in the same area. Unable to contain his growing admiration, the young resident finally spoke up: "I get really distressed," he said, "because there are so many urgent cases, and so often I feel completely buffaloed. Then I look at you, and whatever comes up, you *always* know what to do." Smiling sympathetically, the older physician replied: "Well, son, fact is, I don't always know what to do…but I'm never in doubt."

One of the things I like about that story is the way it humorously highlights a basic dilemma in our lives. We're constantly called upon to make choices—some of them crucial—yet we seldom have enough knowledge to take action with absolute confidence. I think, in fact, that this problem accounts for the attraction of Fundamentalism in all its forms: "The Institution to which we belong," so the story goes, "provides the certainty we desperately need in order to wend our way successfully through life."

It also accounts for a downside of such Institutions. When 'we' all think the same way, it's natural—being members of a highly complex social species—to think we're right. But if some of those around us don't share our convictions, we may start to have doubts. One way of dealing with that problem is to ghettoize ourselves so we don't have to hear the message of outsiders. But it's even better if we can get the outsiders to come around to our view. That being the case, when Fundamentalists gain sufficient power, they're often tempted to *demand* adherence on the part of all comers, justifying their stance with the famous dictum that "Error has no rights."

In marked contrast, a scientific worldview doesn't even aspire to the achievement of absolute truth; and congruent with *that* view, the thesis rendered in the last few chapters makes no pretext of representing reality *exactly* as it is. The explanation's attractiveness will vary in proportion to how much sense it makes to readers, and its scientific usefulness will depend on the facility with which confirming (and *disconfirming*) experiments can be derived therefrom.[1]

SKINNER'S THEORY REVISITED

Within that context then, let's review first the manner in which our present conceptual model differs from Skinner's original *Theory of Operant Conditioning*. His model, you may recall, insisted that the term 'reinforcement' be limited to externally observable reinforcing events. We claimed instead that the time for appropriately viewing the brain as an inscrutable "black box" had passed, and we illustrated the point from an animal study going back to almost the middle of last century. Our specific example involved an experiment wherein a rat's behavior was vigorously reinforced by direct stimulation of a so-called reward center *within* its brain.

Since that time, scientists have continued to extend our understanding of the nervous system, unveiling along the way additional elements of our internal-reinforcement apparatus. For instance, we know that the use of cocaine becomes so utterly compelling to habitues because it intensifies the effect of the neurotransmitter, *dopamine,* in an area of upper brain stem known as the *nucleus accumbens.* And on the aversive side, many external reinforcements work their negative magic through an area of the temporal lobe called the *amygdala* (Latin for *almond,* describing the shape of that highly complex cluster of cells).

The recent miracle of *neuroimaging,* by lighting up the relative activity of various brain areas, has provided all sorts of additional opportunity to correlate cerebral activity with particular modes of experience. For instance, as I write these very words (2003), yet another study has just been published in the *Journal of Neuroscience* detailing an experiment performed on human volunteers called upon to select from a menu of food items. Here were some of the results: Certain areas within the amygdala became more active in proportion to how highly a given dish was rated by individual participants; part of the prefrontal cortex (medial orbitofrontal) became involved in integrating the valuations; and another area in the prefrontal cortex (lateral orbitofrontal) became active when choosing between two dishes of closely weighted value.

But despite such ongoing studies, we still have quite a way to go before arriving at a functional brain map detailed enough to adequately specify internal reinforcements at a neurological level. That's the reason we pitched our explanatory story at what I referred to earlier as the *"motivational level"* of pleasure/pain experience.

Having done so, however, we are now in a position to see even more clearly why Skinner insisted in his original model that only external reinforcements be used. Recall that externally directed observations are relatively straightforward (Chapter 16). If, for instance, a rat's action of depressing a bar is shortly followed by the delivery of a food pellet into its feeding tray, not many critics are about to contest the fact of this sequence.

Let's contrast this situation with the motivational level of explanation. We end up having to say things like *"Look within yourself* and observe: When you treat someone unfairly by your own lights, you'll feel guilty; when you treat someone well, you'll feel pleased." (That's of course in uncomplicated cases; if you treat someone well, and that person then sneers at you for doing so, your feeling of pleasure will likely vanish in a New York minute.)

But note right away the considerable contrast between this procedure and, say, examining a rat's feeding tray. A whole bunch of us can stare at the tray and see if indeed the food reinforcement arrived as stipulated. But no disinterested third party can look *directly* inside the other person's experience to see if the pleasure/pain reinforcement actually arrived. We're reduced to asking that person to *"Look within yourself."* To wit, we're asking that person to engage in *introspection*, the very approach that fostered behaviorism in the first place because *"introspectionism"* led to so many disputed results.

So here we are with *deja vu* all over again, as Yogi Berra might've put it, because others may respond to our introspective request as follows: "Well, *you* may look within yourself and notice that you feel guilty when you treat someone unfairly, and *you* may feel pleased when you treat someone well, but *we* note no such consistent feelings within ourselves. We simply try to treat people fairly because it's the right thing to do. Feelings enter into the equation only incidentally, if at all."

So what's going on here? Am I fancying that I feel something that I'm not actually feeling? Or are the others "not in touch with their feelings?" Or is it the case of different strokes for different folks, some of us responding in pleasure/

pain mode, others of us not? And if the latter is so, would it not prove that moral choices can and *do* operate independently of any pleasure/pain apparatus?

THE VAGARIES OF INTROSPECTION

First point: one of the problems with introspection is that we so often end up seeing what we want to see, and not seeing what we don't care to see. Every psychiatrist, for instance, comes across example after example in which folks seem out of touch with their feelings. Here's an instance that impressed me many years ago: I was attending a devout nun who suffered from acute anxiety attacks ('panic attacks' in current jargon). Although such attacks often come 'spontaneously', unleashed perhaps by a trigger-happy sympathetic nervous system, symbolically laden events sometimes precipitate such attacks also. That's why it makes sense to ask what was going on when an attack started. In happy memory, I struck gold the first time I asked this patient, because from her description it seemed clear that another nun had been harassing her right at the time of the event.

So naturally I asked: "Well, weren't you *angry* at her?"

"Oh, *no*," said my patient.

"How come?"

She smiled at me benignly: "I love *all* of God's children, and I'd never disappoint God by getting angry." We had an interesting discussion then about the virtue of "turning the other cheek," as she put it, but she wasn't quite ready at that point to see evidence of her own semi 'sinful' response.

I gave myself the following explanation: My patient had indeed started to feel angry at her fellow nun, but this nascent emotion was exceedingly unacceptable to her—to the point where it would've totally hammered her self esteem—and it was this danger that had frightened her. But being unable to identify the quickly buried source of her fear, she experienced only an unnamed dread.

This particular instance had a happy ending. Once we had a chance to *acceptably* clarify the nature of such recurrent happenings—one small bit at a time—my patient developed a more reasonable attitude toward her own angry feelings, with the result that such emotions became less 'dangerous', hence less evocative of a fear response. (I'd be remiss not to add that anti anxiety medicine also helped greatly by calming her trigger-happy autonomic system at a more basic level.) But at any rate, the reason I relate this story now is to illustrate that when something's a self-esteem crusher, we often tend *not* to see it.

With this introspective wrinkle in mind, we might claim that those who aren't aware of guilt feelings when they're tempted to do something that's unfair (by their own lights), or who aren't aware of pleasurable feelings when they do something nice, have become so committed to the notion of doing-right-because-it's-the-right-thing-to-do that they'd feel diminished if they opened their introspective eyes to see what's actually the case. They in turn, of course, could use the same genre of argument, concluding that my own self-esteem is on the line as I'm preaching about a favored theory; consequently, that I end up fancying myself into feelings my thesis claims are present. Good way to start a name calling brouhaha, but not much of a way to make rational progress.

THE IMPACT OF HABIT ON OUR AWARENESS

Second point: While we're going to focus on the enigma of consciousness in Part III of this book, it may be useful to anticipate our discussion of that subject a bit now, enough at least to note that explicit awareness of habitually performed acts tends to diminish, sometimes markedly. Take driving, for instance. While beginners will be acutely aware of how they depress the accelerator, clutch, and brake pedals, or just how much they're turning the steering wheel, experienced drivers will usually perform all these same actions without giving them a thought. Same thing applies to our many minor moral choices. Only the 'biggies' tend to make it through, agonizingly, to full consciousness.

Example: Thanks to the beneficence of modern medicine, many elderly parents have been enabled to live *well beyond* the point of mental or physical incapacitation. Their middle-aged children almost always feel guilty while arranging for nursing-home care—along the lines of "Good ole Mom took care of me when I was growing up; it's only fair that I should've taken care of her when it got to be my turn." So then the afflicted children will often quite consciously weigh in with their contravening self-interest program, as in "It's not like the so-called good ole days, when death would come knocking at the door shortly. With antibiotics to ward off pneumonia [what used to be referred to as "the old man's friend"], and with all sorts of other medical wonders, Mom could live for a dozen years or more like this; and I can hardly lift her anyway, even if I could stay awake and on duty for the 24/7 care that she needs; and besides, I do deserve a life of my own too." And so on.

We're all familiar, that is, with the explicit agonizing that comes with landmark instances of moral choice (another one: "Well, yes, I did agree to stand by my man, but..."). However, in the more habitual minor cases, we generally make our choices "without a thought," that is, without consciously attending to the complex of factors that underwrite such daily decisions.

LOOKING AT 'RATIONAL' MOTIVATION

Third point: Those who judge that they try to do the right thing simply because it *is* the right thing hardly see their moral actions as unmotivated. So, for instance, when I talked about Albert Schweitzer in the last chapter, and noted that "a person's life of steady devotion can hardly be viewed as a series of unmotivated (random) acts," they would agree fully with this premise, but not with my proposed version of that good man's motivation. They would be likely to say instead that he performed his beneficent deeds as a result of *rational* motivation.

Problem is, the word *'rational'* has so many different uses that we could've employed it as a banner instance of the imprecision of our "conceptual containers" (Chapter 12). Look in a dictionary now and you'll probably find it defined as *'reasonable'*. Look up 'reasonable', and in turn you'll find its synonym, 'rational'. And though *'logical'* will usually be provided as still another sense of the term, it's clear that more is required than conformity with logic for us to judge that a person is thinking and behaving rationally.

To illustrate: A patient of mine developed the belief that some person or agency had posted observers to watch her comings and goings at my office. How could she tell? Sometimes by the expression on an observer's face. And even more intrusively, she was sometimes being followed by other cars on the highway home. Whose car? Different cars were being used, but she could always note the particular one that stayed discretely on her tail. Who was responsible? Of that she wasn't certain, nor was she certain how they were going to use the information obtained. But she was sure that this *"invasion of privacy"* was being done with evil intent, even though she couldn't be sure exactly what evil might eventually happen.

I made the...*uh*...surprising judgment that my patient's fears were not "rational." But I wouldn't be able to demonstrate any strictly logical inconsistency in her thinking. For instance, she never contradicted herself. Our differing conclusions about the situation came, not as a matter of pure logic, but from the fact that two of our information-processing routines *weighted* matters so differently.

Routine one: We have a built-in alarm system that helps us smell out incipient dangers—prominently involving hazards from others of our own species, which is hardly surprising since human predators will be lots cleverer than the others (accounting for the old expression, *"homo, lupus hominis,"* or "man, the wolf of man"). My patient's early-warning system caused her to be as hyper alert in this regard as a sentry trying to read shadows in the dark. Alternately stated, her system provided very heavy *weighting*—way too much—to any prodromal sign of possible danger.

Routine two: We have a built-in program that automatically draws a proportion between the strength of people's motivation and the amount of energy they will exert (though as is the case with all our complex programs, this one gets fleshed out by personal experience and learned stereotypes). My patient's program lacked the exceedingly heavy *weighting* of my own in this respect. From my perspective, that is, unless someone were sufficiently motivated to spend the many thousands of dollars that would've been necessary to keep tabs on her visits in the manner she feared, it wasn't actually about to happen.

Point of our illustration: When we judge that a person is thinking *irrationally,* it's more often because their weighting systems are out of kilter, rather than that they're violating strict principles of logic. Same in the other direction. When we judge that folks are thinking rationally, we accept that they're thinking *logically*—though oftentimes in fact our putatively rational thinking is *not* totally logical—but we also accept, at least tacitly, that their *weighting* routines are at least in the right ballpark.

Back to specifically moral choices, people who emphasize "rational motivation" usually link their thesis with a chain of reasoning that flows *logically* from some basic premise like "obeying God's law," or adhering to a "categorical imperative." Their chain of reasoning, in turn, usually features the concept of *'duty'* (for instance, "When God gives a command, it's our duty to obey"). Note, however, that 'duty' is not simply a factual concept. Those who feel a duty to do something feel guilty if they don't do what they judge they should do, *and* they feel a sense of satisfaction when they fulfill their duty.

MEASURING PLEASURE/PAIN SYSTEMS OBJECTIVELY

Lets' return now to the earlier remark that our "explanation's scientific usefulness will depend on the facility with which confirming (and *disconfirming*) experiments can be derived therefrom." Thanks to the miracle of modern technologies like neuroimaging, it will become possible to devise studies assessing the validity of our Hedonic model. That's because the motivational pathways it requires would be activated in all subjects tested, whatever their experiential reports might be. In turn, this activation will be rendered as a brain representation (usually implemented by varying light intensities or colors) that anyone can inspect to see if indeed the internal reinforcements arrived as stipulated.

ONE MORE OBJECTION...

There's a final objection of great historical importance—one we might refer to as the *Katie-bar-the-door* concern—so let's take a look next at this difficulty.

19

KATIE BAR THE DOOR

Bertrand Russell once noted that no one to his knowledge had ever been burned at the stake for denying our most evident truths like, say, 2 + 2 = 4. Being toasted alive seems to have been reserved for those inclined to deny more recondite notions. And while we may not have made ideal progress when it comes to the toleration of dissident thinking, we have improved our record a bit lately in comparison to the good ole days.

For his time though, the great John Locke (17th century) was a very tolerant thinker. As an example, he was up to the task of accepting different ways in which the Lord might be worshiped. But for all that, he could not abide atheists, for as he put it: *"The taking away of God, even only in thought, dissolves all."*[1] What was he thinking about there? Why didn't he just treat those who denied 'God' as benign simpletons? That's presumably the way he would've looked at those who might deny '2 + 2 = 4'.

Here was his problem: If one looked at the situation from a viewpoint common at the time—and one that's still pretty common—God had been assigned two crucial tasks by humankind: **1.** He was to be the guarantor of our continued life after death, and **2.** He was to mete out perfect justice, making up to all of us latter-day Jobs for the unfairnesses of our present lives—as well as seeing to it that all those scoundrels thriving amongst us end up getting their just deserts. (Since **'2'** obviously doesn't happen in this life with sufficient regularity, we've obviously got to have **'1'**, which is a good thing anyway, given the fact that we're normally programmed to fear death and to want to go on living. Indefinitely.)

Within such a context, doing away with the notion of God as policeman/judge would—so the intuition went—give folks free rein to do anything they felt like. That being the case, our personal security would be placed in extreme jeopardy. We'd be reduced to the most desperate measures in order to save life and limb; hence, *"Katie bar the door!"*

Though theorists were of course familiar with the notion of one's "voice of conscience," this phrase was usually, and I suppose aptly, supplied with a crucial modifying adjective, as in "*wee* voice." No one, that is, seemed too impressed with the robustness of this built-in mechanism. However, add a policeman God to look over everyone's shoulder, and there'd be a workable mechanism to keep folks at least somewhat in line.

THE UNDERLYING PREMISE

Let's take a moment now to state explicitly the premise underlying this judgment, in order to see if the idea merits acceptance as one of our enduring truths. Here's how we might put it into words: *"Without the colossal carrot and stick of heaven and hell, people will wreak havoc on their neighbors."* On second thought, since some 'God abiding' people are already doing a pretty good job in that direction, let's add a quantitative qualification: *"Without the colossal carrot and stick of heaven and hell, people will wreak <u>even more</u> havoc on their neighbors."*

If we were to examine that thesis before the fact, behaviorists might well express doubts about its likely validity. The problem has to do with timing. Recall that pleasures and pains that happen right away so often trump pleasures and pains that recede into the indefinite future. Alternately stated, the reinforcement that gets there first tends to win out. Those arguing the case from the other side might of course counter that the *colossal* carrot and stick created by literal belief in heaven and hell would be enough to carry the day—especially given our species' ability to create *virtual realities* and relate them vividly to any current situation.

EMPIRICISM vs. REASONING FROM THE EASY CHAIR

The scientific worldview takes such theoretical disagreements in stride because there are very few instances in which one could *not* muster arguments favoring a given thesis over its rivals. Scientists are wedded, however, to the notion of looking to see what's actually the case, ideally under conditions where major variables can be controlled. Unfortunately, I'm not aware of any formal experiments having been ventured in this area, so the best we can do at present is to look at the 'open studies' provided by personal experience.

My own observations surprised me initially. That's because I was raised in a religious tradition that considered the reinforcements of heaven and hell (we called them 'sanctions') absolutely crucial in promoting moral behavior. Naturally then, that's what I was expecting to see. But what I *actually* saw as I went along in life was that those who professed no belief in God seemed as a group to be no worse (or no better) than those who believed in a rewarding and punishing deity.

Intolerance of 'atheism' then was based primarily on the intuitive notion that to deny the existence of a Personal God would lead to a total breakdown of society (*"dissolves all"*), because society would come apart without the accompanying carrot and stick of heaven and hell. Yet in our current "post Christian era," where so many of our citizens are not subject to belief in these sanctions, society has not collapsed. And in fact, given the way things are going currently, we might even do well to add a caveat for some of those who *do* believe in the ultimate reinforcements of heaven and hell, like "God may not reward you with heaven just because you blow yourself to bits, in order to take down a bunch of other folks who are stiff-necked enough not to accept the sacred truths you happen to know concerning Him and His ways."

"VOICE OF CONSCIENCE" HAS A DOUBLE ASPECT

A little while ago, we placed the modifier "wee" next to the phrase "voice of conscience," and we did so because our sense-of-fairness *does* so often strike us as underpowered. Making its lack of vigor seem even worse, however, people don't tend to give enough thought to the positive half of this internal-reinforcement system. Reminds me now of a story I came across years ago concerning a New York bus driver. Seems he'd read about an old lady who left a big stash after her death to a driver who'd done her a good deed. So he said to himself: "This'll be like a lottery for me. I'm gonna' treat folks real nice on my bus, and maybe someone'll remember *me* in their will." With that in mind, he started going out of his way to help the elderly and infirm on and off his bus...and so on. But then he began to notice something he wasn't prepared for. He was actually *enjoying* his work, whereas before it'd been just a job. I believe this self-report was probably accurate, because over the years a number of my patients and social acquaintances have provided me with similar accounts, suggesting a rather robust presence of internal reinforcements on the *positive* side.

BUTTRESSING THE "VOICE OF CONSCIENCE"

Nevertheless, when it comes to keeping people on the straight and narrow, it's still common to think first and foremost of negative reinforcements, and it's specifically the reinforcement of guilt that we usually think of as insufficiently motivating. Probably this judgment would be on target if the guilt experience were not additionally buttressed by the *virtual reality* provided by external reinforcements. For instance, most citizens think it's *fair* to pay their share of running our Country. But given the many other calls on their money, most people are happy to shade the amount of income tax they end up paying. Any guilt feelings tend to get successfully countered by a number of factors, including knowledge that our tax system is, practically speaking, not based so much on fairness as on compromises hammered out by pressure groups that can exert the most power. Nevertheless, the possibility of criminal repercussions keeps most people from trying to shade things too far, because the *virtual reality* of going to jail acts as an effective *external* reinforcement, adding potently to the impetus of one's inner "voice of conscience."

INTERNAL REINFORCEMENT OF *EXTERNAL* LAWS

All that said, our inner reinforcements probably *are* stronger than we usually give them credit for. One way of checking this out is to observe how much better compliance becomes when a civil law echoes our inner sense of fairness. Take the common traffic law requiring drivers to stop in either direction when a school bus is flashing its signals to take on or let off school children. The cumulative delay, if there's no place to pass, can be a real pain in the butt. Nevertheless, in all my years of driving I have yet to see anyone violate this law. Why? Probably because there's such a consensus that it's only *fair* to protect children, because they're our future and they may yet be too impulsive to attend optimally to traffic. Contrast this law abiding behavior with the fact that it's hard to find anyone who *can* be counted on to obey speeding laws. Why? Probably because we tend to have thoughts like "a 55 MPH limit doesn't make sense along this stretch, at least for a good driver like me." (And by the way, I've yet to meet anyone who thinks of himself as a bad driver.)

MURPHY'S LAW

An additional way of marking the actual strength of our internal reinforcement systems is to observe what happens when such motivating mechanisms are missing-in-action, so let's look now at moral behavior when Murphy's Law steps up to the plate: *"What can go wrong will go wrong"* is usually employed as a humorously pessimistic aphorism. Also turns out, however, to be a serious engineering principle. Suppose, for instance, that the chances of an engine failing are no more than one in a *billion* cycles. With odds that great, who's to worry? Yet suppose this engine cycles at 5000 rpm while it's actively operating over a 10 year period. If we do the arithmetic, it's apparent that the engine is likely to fail more than 25 times during this interval, so we'd better have contingency plans set up in that regard.

Point I'm getting at is that *all* engineering programs have a certain rate of failure, and biological programs are hardly an exception.[2] Given a combination of genetic coding errors, developmental glitches, and environmental forces, it's in fact not at all rare for this to happen. And when the subtle programs we've been talking about don't develop properly, affected individuals lack the normal tendency to feel guilty when they treat others unfairly, and they lack the normal feelings of pleasure when they do something nice for their friends. Consequently, they can respond only in terms of external reinforcements like the danger of incarceration. Perhaps not surprisingly, the latter reinforcement of itself is often not up to getting the job done, accounting for the fact that people with *Antisocial Personality Disorder* (or *Psychopaths*, as they used to be called) are highly over represented within our penal system. Additionally, lacking the positive reinforcement of pleasure when they treat others well, they're also in tough shape when it comes to normal affiliative bonding.

To get the feel of what this deficit is like, try to imagine for a moment that you suffer no pangs of guilt when presented with the opportunity of taking advantage of another person, say, your significant other. Under such conditions, whatever you do or don't do, you're not going to feel guilty about it. Further, you won't miss the pleasure you experience when you do something nice for your loved one, because you never experience that particular kind of pleasure anyway.

You would have one advantage though, because it'd be much easier to focus on your own narrow self-interest, unfettered by the pleasure of pleasing others, and unpained by taking advantage of them. Seems like a good deal in the short run. Of course, it may not surprise you to hear that one of the typical findings with psychopaths is their lack of close relationships. (Apparently, it's hard to find many folks willing to put up with psychopathic behavior on a regular basis.)

But there's also something to be said for this failed equipment, if you're in business. For whether you're a salesman pitching to customers, or a C.E.O. trying to sell shareholders and the public, it's much easier to con folks if you don't suffer from scruples that are likely to give you away. Of course, you would have to keep moving on, because too many people seem fond of implementing that old aphorism: *"Fool me once, shame on you; fool me twice, shame on me."* Not surprisingly then, it's a characteristic pattern for psychopaths to move around frequently—new work venues, new social groups, and new geographical locations.

Two more points before leaving the subject. First, it's important to make a distinction between those who lack the equipment needed to develop an appropriate sense-of-fairness, and those who have the equipment but who learn an aberrant language-of-fairness. Pursuing the language analogy, true psychopaths are akin to those born with congenital aphasia (lacking the equipment needed to learn language), while common criminals are more akin to those who learn bad grammar. An example from the latter grouping: Mafiosos have a strong code of honor when dealing with those within their own "Family," but outsiders are fair game.

The distinction between psychopath and common criminal is important for a second reason. Since the average intelligence of psychopaths is less than normal (IQ median around 90 instead of 100), they commonly trip themselves right into the penal system where they are generously represented. In marked contrast, however, many of the highly intelligent ones are clever enough to stay within a grayish area that shrewdly pushes existing laws. These individuals are often able to benefit handsomely from their lack of scruples, while for the most part avoiding any effective prosecution.

The number of psychopaths in the community is discouragingly large (roughly 2 % of the total population), and the damage they wreak on the community is costly indeed. For all that, our present discussion has only the limited goal of emphasizing problems that arise when our human sense-of-fairness program is not up and running. If there'd been a 10 fold increase in the prevalence of psychopaths (say, 22% of the population), it's unlikely that our species would still be hanging around. And this aberrancy highlights the crucial difference that our normal *sense-of-fairness* makes when it comes to the way in which we conduct our human affairs.

20

POSTSCRIPT TO PART II

When we're talking about our most sophisticated weighting programs, we usually employ the term 'values'; and in PART II we've centered our discussion on one such program, namely the sense-of-fairness program—a fitting choice in so far as the moral sense constitutes our most crucial value system when it comes to dealing with each other. Restated for greater emphasis, no highly sophisticated social species like ours could make it through even a single cycle of reproduction without a sense-of-fairness.

However, our moral sense is by no means our only vital value program. Our sense of beauty—our esthetic sense—may not be too far behind in importance, because the pleasure it provides 'helps us bear those ills we have, rather than rushing to others that we know not of'. So before moving on to PART III, I'd like to mention an instance of esthetic weighting, in order to provide at least some suggestion of the dovetailing that occurs among biological programs.

MOTHER AND CHILD REUNION

I mentioned earlier that visitors to my homestead have always tended to *"ooh"* and *"aah"* when they behold new lambs, especially as affectionate interaction unfolds between lambs and their mothers. Observing new life seems to be a renewing experience for all of us in fact, and observing the caring ways of a mother as she attends to her newborn evokes in us a pleasurable warmth that no doubt hearkens all the way back to our early mammalian ancestry. Scant wonder then that artistic depictions of mother & child have been perennially popular.

Within our Western tradition, this theme has commonly been portrayed by way of the *Madonna and Child* (Mary and her God Child). To Christians, the visual impact becomes awesomely enhanced by literal belief that a member of our own species has been selected as the "Mother of God," enfolding the very Author of our universe within her womb and then within the comforting cradle of her arms.

Given such a scenario, it's hardly surprising that the faces of both mother and child are so often highlighted by halos of golden light. And all the more joy arises in true Believers, because Mary's Son is soon to restore Divine light to our world. Fittingly then, this image of human renewal ('redemption') is festively celebrated immediately following the winter solstice, at a time when light is literally return-ing to our world (Northern Hemispherically speaking).

Appreciation of a scenario of this kind comes naturally, however, not only to Christians, but to all of us as self-conscious mammals. That's due to the fact that our taste in beauty flows so directly from our biologically derived weightings. We dearly appreciate our mothers, without whom we would not exist. And all moth-ers bear a basic relation to the matter of our own earth, which in turn is related to a more Primal Source. (Congruently, the Latin word for *mother* is 'mater', derived from the same Sanskrit root that gave rise to words like 'matter', 'mare'...*and* 'Mary'.) At a gut level then, *all* members of our species can respond positively to the Yuletide story and its various depictions, even if they don't accept the story as gospel (just as those who lack literal belief in the ancient myths can nevertheless relate meaningfully to stories about Isis or Athena).[1]

WEIGHTING AND LOGIC

Let's move now from this esthetic scene back to our sense-of-fairness program by imagining the following situation: The mayor of Santa Town U.S.A decides to place a creche in front of the Town Hall, along with the statues of Santa and his reindeer that are already present. Some local citizens, however, take exception to this bold statement of Christian doctrine that's being foisted upon them at their very own Town Hall. So they enlist legal help to protest "this obvious intrusion of Religion into secular government."

The mayor replies that he's not a member of *any* Institutional Religion. He claims moreover that the Christian pageant, in his view, is no more and no less than a story of breathtaking beauty, set upon a background of human failures and disappointments (symbolized in this "ancient myth," as he calls it, by Adam and Eve's 'Fall' and mankind's ongoing 'sinfulness'). But the story encourages and elevates us, he insists, because it provides us with a magnificent symbol of human renewal, the God Child's birth—not just any new life after all, but a baby who is human like ourselves, yet possessed also of that spark of Divinity toward which we symbolically aspire. The 'Lord Jesus' is then, from the mayor's perspective, a great mythical figure (albeit based on an actual human being) who symbolically 'redeems' us despite our past failures, by providing the 'grace' we need to overcome our flawed limits. And how fitting, he insists, that we should celebrate this most beautiful of stories each year, just as the sun's light is beginning to return more strongly to our world. Indeed, what better time could there be to salute the prospects of a new beginning? What better time could there be to honor the brotherhood of man by bestowing gifts upon each other—just as the wise men stationed outside the creche bear gifts for the God Child, who in turn symbolically gifts us with the prospects of 'salvation' from our troubles and our deficiencies.

Nonsense, the lawyers retort, this sort of argument amounts to no more than a thin rationalization aimed at unseparating Church and State. Nonsense back, says the mayor, I'm not even a Christian, and if we're not allowed to portray the Christmas Story at our Town Hall, we would in like manner have to dispose of the symbols of Santa Claus, who is none other than the Saint Nicholas of Christian tradition.

And, he continues, what about the Easter Egg hunt we hold at the Town Hall each Spring? Will we have to cancel that celebration too? After all, this delightful pageant is tied to yet another religious feast, the celebration of Christ's Resurrection at Easter time—an event that heralds for Christians their very real resurrection from the dead into an eternal abundance of life. But the pageant of an *Easter* Egg hunt does no more really than to symbolize our natural human quest for more abundant life, as does the Easter bunny (given the fact that rabbits are fabled for their fecundity), and as does the meaning of Easter itself. Many of us, that is, take *all* these stories as powerful human symbols, and how fitting that we celebrate the feast of Easter in early Spring, at a time when new life is awakening all around us.

The lawyers respond, in turn, that the purported analogy between the Christmas portrayal and an Easter egg hunt—or any claim that the evocation of Santa Claus (*aka* Saint Nicholas) amounts to the expression of Christian doctrine—is absurd. Santa Claus, Easter eggs, and Easter bunnies, they maintain, are obviously far removed from the teachings of Christianity, whereas the creche involves a direct, visually expressed statement of Christian doctrine.

So how would a judge adjudicate the issue? How, in other words, would he weight the various arguments pro and con in order to arrive at a fair decision? Or might he contend that by applying *pure logic* to the initial premises provided by legal precedents, he could reason directly to a proper finding? And if he did indeed hold to such a narrow notion of what makes up a rational ruling, would we not be tempted to remind him of that humorous old saying: *"Logic is a systematic method of coming to the wrong conclusion with confidence."* For logic, if it's to produce useful results in our ordinary world, must do its work within the framework of our human *weightings*. Otherwise, its employment becomes, paradoxically, an irrational exercise.

PART III
THE ENIGMA OF HUMAN CONSCIOUSNESS

INTRODUCTION TO PART III

Consciousness is at the very core of our human existence—so much so that most of us wouldn't care to have our bodies kept alive if the areas of our brains underwriting this wondrous capacity were destroyed. For us to be *humanly alive,* that is, we need to register our world consciously, then be able to think and feel about it.

And though we're clear at this point in human history on the *fact* that a functioning brain is necessary for human consciousness, how this amazing attribute actually arises remains mysterious. For how can it be that a three pound cluster of corrugated flesh occasions such an astounding ability? Indeed it's almost weird, when you think about it, that the most immediate and most intimate aspect of our own selves remains so difficult to fathom.

Fortunately, most sensible folks don't waste a whole bunch of effort agonizing over this enigma, usually thinking about the issue—and then only indirectly—on occasions like the death of a loved one, a time when most of us are quite happy to approve the *traditional explanation* (one that's been nourished from as far back as we have human records), because its acceptance provides at least some comfort as we're trying our best to deal with the horrendous loss we've just experienced.

And we should probably have put *"traditional explanation"* in the plural, because there are many variations to the story. In all cases though, the <u>conscious self</u> continues to exist after its container dies. Implied is the notion of a <u>spirit substance</u> that can detach itself from the body—a special sort of substance that's responsible for our consciousness. But the problem with this 'ghost' account is that its various versions fall into logical inconsistency when we take the time to look at them more closely—which we will do.

Because of the intractable problems that arise in attempting to provide a reasonable account of the interaction required between a separable <u>consciousness substance</u> and its body, almost all modern theorists-of-mind have given up this explanatory framework (known as *'Substance Dualism'* because two separate substances are invoked). Instead, current theorists work from a basic notion that the way our brains function—most especially their multilevel information processing activity—somehow gives rise to consciousness.

We'll address one of the commonest versions of this view, what we might call (echoing the terminology of particle physicists) the current *"Standard Model"* of consciousness. Only problem with this model, as we shall see, is that it doesn't work very well either. During our journey, we'll also see why some modern theorists of mind have been keen on getting rid of the whole notion of consciousness (or as their critics are wont to say, why such thinkers end up "feigning anesthesia").

A PERSONAL VIEW OF CONSCIOUSNESS

When it comes to my own explanatory turn at bat, I probably won't do much better, but while I'm giving it a go, readers may well notice some similarity between me and the older physician who was working at the hospital emergency ward. You remember the attitude he brought to his job: "Well, son, fact is, I don't always know what to do…but I'm never in doubt."

What I *am* planning to do—with utmost certainty—derives from my admiration of modern particle physicists; hence, I'm going to follow their strategy. Whenever they come across something so basic that they can't break it down into more fundamental constituents, they simply accept it as a *given*. Force, for instance, is a given. Recall that the authors of the *Encyclopedia of Physics* didn't even provide an entry. Problem of course is that the notion of force seems too fundamental to allow formal definition; yet its effects on us are so compelling that to deny its existence in our *ordinary world* would amount to no more than intellectual posturing.

Consciousness also defies formal definition, since it can't be broken down into more fundamental constituents either. That being the case, I take the approach: *"Consciousness is another one of those fundamental items that's there all along."* The good news is that I don't even have to invent a name for such a theory, because there's been a label hanging around for centuries. Bad news is that the label, *panpsychism,* involves a notion that's also been pooh-poohed for centuries. We'll get to the reasons why in due course. But under the circumstances, I'm going to need some fancy footwork if readers are to take my conceptual model of consciousness seriously.

Another thing we'll need to do along the way is distinguish carefully between the notion of *Naturalism*—the explanatory framework we adhere to exclusively in this book—and *Supernaturalism*. This distinction will become particularly important since even the most famous of traditional explanations for consciousness, the one supplied by the great Descartes during the Seventeenth century, requires a Supernaturalist world view. We'll start our journey now by looking at some of the traditional theories.

21

THEORIES OF CONSCIOUSNESS FROM DAYS OF YORE

The earliest evidence we have of humans developing a theory of life-after-death comes from archeological unearthings of apparent burials. For instance, one site in France from about 70,000 years ago contained the remains of a young Neanderthal man (*Homo sapiens neanderthalensis*) who had some tools and food placed alongside. First thing to remark on is that no other mammals—including earlier hominids—have engaged in burial behavior. And even if burials were performed for practical reasons like hygiene or so as not to encourage visits by predators, why would our forerunners have wasted food and valuable tools by burying such items with the deceased person?

It was hard for anthropologists *not* to speculate that the dead man was probably dear to his fellows, who thought that food and tools would still be useful to their comrade in some way. If this speculation hits anywhere near the actual target, then it follows that his fellow Neanderthallers held a belief in life after death—which in turn implies that some aspect of the young man was to be separable from his obviously dead body; and further, that this 'whatever' must be made of something much finer than the ordinary stuff of his decomposing carcass.

WE ARE ALL PRACTICING DUALISTS

Before examining such 'ghost' substances, however, we should note that the way we experience ourselves does much to promote a division into two separate entities, body *&* spirit. That's because we naturally think of ourselves as quite distinct from the collection-of-body-parts that make us up. And we also introduce such a distinction into our everyday language, where we routinely state a proprietary relationship between this self and its body. To wit, we say "I am myself," whereas "I have arms and legs." In similar fashion, each of us feels like the captain of his ship, not only owning his various body parts, but responsible for directing them in the appropriate manner (as when a tennis player says "I've got to keep my eye on the ball").

And there's another factor: Our cognitive equipment requires us to think in terms of *things;* we only awkwardly think in the direction of *not* things. In fact, it's impossible for us to have an active thought as to what our *not being* would be like, for how does one have a positive thought about a *no* thing? Hence it's easy to think of oneself as existing come what may, but impossible to think directly of no longer existing. The Dualist manner of thinking arises naturally then, given our kind of self-conscious cognitive apparatus—so much so that modern theorists had to overcome the strong impact of our everyday way of thinking before finally arriving at the current consensus that <u>our sense of a conscious self *separable* from its body is an illusion of human experience rather than the actual reality</u>.

Since some readers may express surprise at this consensus, let me add a couple of qualifications: First, I'm not including "story tellers" in the category of serious theorists (as will be illustrated below), nor am I including theorists who work within Institutions *requiring* acceptance of Substance Dualism as an unquestioning matter of Faith. We will, as we progress, have occasion, however, to remind Christian readers that it was not Jesus who introduced the notion of two separate substances, but rather some of the Platonically influenced theologians of ancient times. And we will note further that scientifically sophisticated Christians like the Anglican priest/physicist, John Polkinghorne, have also rejected this logically untenable notion, even though they look forward as avidly as Christians of yore to *"the resurrection of the body, and life everlasting, amen."*

WE *WANT* OUR CONSCIOUSNESS TO OUTLIVE OUR BODIES

What made the current consensus of theorists even more difficult to arrive at is the fact that we so dearly *want* ourselves to outlive our all too mortal bodies—a powerful urge indeed, resulting from crucial biological programs that make us dread the prospects of our death and yearn for ever more life. Without such programs after all, who amongst us "would bear the whips and scorns of time?" Yet once we relinquish the notion of a separable <u>consciousness substance</u>, we have to reconcile ourselves to the definitive ending of our lives at the time of our deaths.

Unless, that is, some External Agent were to provide operational replacement of our bodies (exactly what up-to-date Christians like Polkinghorne believe is attainable from a Transcendent God who is involved in human affairs.)

In summary, we start with a self-conscious animal who thinks naturally in Dualist mode (me & my body) and who can't even cogitate directly about himself as *not being*. Then add to the mix our longing-for-life and fear-of-death programs, and the traditional explanatory story becomes difficult to reject. Notwithstanding, it's time for us now to take a closer look at the spirit substance that's been proposed for the purpose of underwriting our consciousness.

SPIRIT SUBSTANCE

Accounts of the spirit substance have varied according to time and place, but all proponents agree that it's markedly different stuff from the ordinary matter making up our bodies and the things around us. The consciousness substance was customarily thought to be something very, very subtle—similar in that respect to the air we breath. And in fact it's no coincidence that the Latin word *'anima'*, so often translated as *'spirit'*, also means *'breath'*. To understand the association, imagine an earlier time when people customarily died at home. With family members looking on, their loved one would heave her last breath, and it was easy to think that this final exhalation was none other than the person's *'spirit'* taking leave of its body.

One didn't usually see the spirit as it departed. But later on this 'ghost' sometimes became visible, at least to some people, under some circumstances. The spirit substance provided its owner with continuing consciousness after death, and in imaginative accounts it also took up space, though often rather amorphously so, as when ghosts are depicted in the form of billowing bed sheets with a couple of eye holes. At other times, however, the spirit substance was envisioned as diaphanously duplicating the exact outline of the body it once inhabited, clothes and all (though...*uh*...the latter aspect may be an artifact of human modesty).

While the origins of such depictions fade into the mists of ancient times, the same manner of thinking remains currently popular, and not only among our New Wave brethren. Let's illustrate the point from a highly successful movie of recent vintage, in which the hero's spirit rises from his murdered body as an exact doppleganger—leaving no member of the audience in doubt as to who this translucent guy is, even though his lifeless body is still lying on the ground. How, one might ask, is his ghost substance able to interact causally with ordinary matter?

Answer: BIG problem. For instance, when our dead hero attempts to visit his beautiful young widow, his hand moves impotently through the knob so he can't open the door. Apparently the *impenetrability of matter* doesn't apply to this spirit stuff. But not to worry, our movie hero does get to see his wife after all, because the same absence of impenetrability allows him (as he soon discovers) to walk straight through the wall.

Before approaching the typical inconsistencies within such a story, it's important to recognize that ghost accounts arise from the poetic side of ourselves, so there's no need for total consistency. A veneer of realism will do—just enough to make the wish-fulfilling story seem more or less plausible. But to illustrate the problems that arise when one tries to take such accounts with full seriousness, let's examine what happens next: We find our dead hero inside his home, where he discovers that his widow is in danger. Understandably concerned, he rushes upstairs to her bedroom. But how is it that though his spirit substance moves right through walls, it interacts in the normal manner with the stair case? Shouldn't his anxiously pounding feet plunge straight through the stairway as he's running up to help his wife?

The story teller can only hope that our concern about the lover's danger will distract us from such considerations. But once we're outside the story's unchained melody, the point is in fact crucial. For if a substance is to interact reliably with ordinary matter, it must possess *predictable* properties that allow it to interact in a *consistent* fashion.

MATTER <u>IS</u> THAT WHICH INTERACTS WITH OTHER MATTER

Our hero's spirit substance in this story *is* made of some kind of matter, given the fact that it's both extended in space (the old time definition of matter), and also that it shows at least some interaction with ordinary matter (the latter property being the ultimate criterion for the modern judgment that *matter/energy* is present).

The problem in the story we're assessing, and in *all* such imaginative accounts, is their inconsistency. That is, when it suits the story teller's purposes, the spirit substance violates the impenetrability of matter, as in our hero walking through walls. Yet when it's convenient for the story's progress, the spirit substance shows the characteristic properties of matter, enabling our hero to pound right up the stairs to his widow's bedroom. (And later on, after considerable 'effort', our dead hero manages to exert an efficacious causal influence on some of the ordinary objects around him.)

But why, one might ask, can we not fashion an explanatory story about spirit substance that *is* consistent, and that we can therefore take with full seriousness? Certainly does seem like a worthwhile project, since if we were successful, we'd be pretty much given a natural guarantee that death would not be the end of our lives.

With that goal in mind, we might note for starters that we do have first rate scientific evidence of stuff that interacts with our ordinary world, yet only *very* weakly. Neutrinos, for instance, can pass right through doors and walls and you and me (and the whole planet for that matter) without necessarily interacting with anything, though occasionally these ghostly particles will interact. So we might speculate that our putative spirit substance is composed of something like neutrinos. It couldn't be neutrinos themselves though, because we know enough about them to recognize that they're not up to organizing themselves into a highly integrated structure that could hold itself together and function in complex thinking, feeling, and perceiving modes as we in fact do. Furthermore, we know the statistical probability of neutrinos interacting with other particles in their vicinity, and it's not even close to enough for them to be able to perform the incessant interactions that a spirit substance would be required to perform in order to control its body. But not just neutrinos, *any* stuff that's only weakly interactive would have the same problem, that is, not being able to perform the dependably continuous interactions required for bodily stewardship. Alternately stated, a substance that interacted vigorously enough with ordinary matter to 'read out' a constant flow of messages from its body and send back a continual flow of instructions, would be anything but weakly interactive.

We'll take a further look at the problem of interaction between a spirit substance and its body when we address the more sophisticated consciousness model proposed by René Descartes in the early 17th century. But let me anticipate a bit now by mentioning that the problems with his model turn out to be at least as intractable.

THE TV WATCHER'S VERSION OF OUR CONSCIOUS SELVES

Let's take up now an additional aspect of the proposed spirit substance, one that *does* make the hypothesis seem attractive on initial viewing. To begin with, recall that we in fact experience ourselves as quite distinct from our collection-of-body-parts. That being so, as we learn more about the function of things like eyes and brains, it becomes congenial for us to think that 'we' view the images constructed by these body parts. In fact, so much was this way of thinking in vogue during my med school days that my teachers used this very model, apparently never taking time to think the matter through.

Little knowledge was available then about the specifics of neural information processing, although of course it was well understood that our eyes detected visual stuff and passed the information along through the optic nerves to visual areas in the thalamus—which presumably did something with the information before sending it to the calcarine cortex in the rear of the brain, which in turn processed and shared its information with vaguely designated "association areas" located more anteriorly. At the end of the line, *voilá,* a fully elaborated image was available for 'us' to view.

Note, by the way, how neatly this model dovetailed with our viewing of the TV sets that had just recently permeated our lives back then. Rabbit-ear antennas (in those primitively privational times) picked up the EMR signals, much as our eyes pick up the visually relevant EMR within *their* bailiwick. The raw data was then appropriately processed by the TV set's innards (vacuum tubes and all) with a picture ultimately projected onto the TV screen, which 'we' were also able to view. In both cases then, the electrical mechanism (TV set or brain) did the detecting and image elaboration, while 'we' did the viewing—all of which reinforced a congenially common sense conclusion that had also long been held; namely, *the viewer needs to be outside what's being viewed.*

I mentioned above that my teachers apparently never took time to think the matter through. That's probably because most people were still *implicit* Dualists at the time (not to mention that anyone like myself who'd been educated by the Jesuits had been *explicitly* trained in this tradition). But let's take a closer look now at the notion that *"the viewer needs to be outside what's being viewed."* Since we certainly do *not* identify ourselves with our brains, this account fits nicely with our first-pass experience. That is, once I become aware that my brain is doing lots of information processing, I then quite readily develop the sense that it is 'I' who views the results of the work done by my eyes and my brain.

Alternately stated, we imagine that this ethereal thing called 'I' (sometimes imagined as a 'little man' or *homunculus* at the control panels) does the viewing. But if we're going to stay within the realm of logic, we can't just pick and choose when we feel like applying a principle. So if the viewer needs to be outside what's being viewed, then this 'I' thing (spirit substance, 'little man', homunculus, whatever) needs in its turn to have a viewer to do the viewing of what it's just come up with when *it* viewed the neural image…and on and on, *ad infinitum*. The only way to break this endless regression of viewers within viewers in a principled fashion is to conclude that the initial premise of a separate 'I' that's needed to view the brain's handiwork, though happily in tune with our human way of experiencing things, just won't square with a logically consistent view of what is. (Let's note in anticipation that we'll be addressing later the issue of what's actually involved in this *sense-of-self* that we refer to as 'I'.)

To conclude then, not only is the existence of a viewer outside our information-processing apparatus not necessary, the notion leads to a logical quagmire populated by an infinite regression of viewers within viewers within viewers. Under the circumstances, our only reasonable option is to quash this homespun notion from the start.

As to why we ordinarily experience our bodies as separate from our 'selves', at least one reason stands out clearly: The human cognitive apparatus has evolved primarily for adapting to our *external* world. (Most responses to the *internal* environment occur at basic levels of adaptation that seldom even make it through to consciousness, like body-heat control, blood-pressure regulation, acid-base balance). And when it comes to dealing with our *external* world, we *are* indeed outside all the objects we perceive. Not surprisingly then, externally directed sensory systems have been adaptively selected to project these objects to positions *outside* ourselves. Hence, when we see or touch our own bodies, we do so by means of our externally directed senses—detection devices that have been preprogrammed to provide externalizing reports.[1]

Let's move on now from our examination of primitive notions of a spirit substance to the more sophisticated model of *non material* substance that was developed by René Descartes.

22

THE SUBSTANCE DUALISM OF DESCARTES

Just as most educated non-physicists are familiar with Einstein's formula, *"E = mc²"* (though they might not know off hand what the 'c' refers to), most non philosophers are at least as familiar with Descartes' famous saying, *"I think, therefore I am,"* though in the latter instance they might not recall exactly what project Descartes was working on at the time. Turns out he wanted to achieve the philosopher's stone of bedrock certainty. In fact, he had as much zeal for this project as any of our current Guardians of the Truth. But not wanting to jolly himself into unwarranted acceptance of bogus truths, he started by methodically doubting everything he could. That is to say, pretty much everything. But *mirabile dictu*, he finally hit upon something he just couldn't doubt, namely his own existence. That's because with all this thinking going on (what with his doubting this, that, and the other thing), there simply had to be some *thing* that was doing the thinking, namely his 'I'. And from this base of certainty, namely his *'I' thing*, he planned to erect an edifice of truth.

It didn't take long for him to go astray, however, because his central bulwark of certainty, *"I think, therefore I am,"* caused problems right from the git-go. To understand what happened, let's examine the progression of Descartes' thinking in slow motion. Recall from PART I that each of our concept/words consists of a verbal object that is *about* something. It's quite easy then to slip into concluding that the "something" the concept is *about* consists of an actual object. For instance, if I look out my window now and see a man at the front gate, and I say to my companion "There's a man at the gate," then my word 'man' is *about* an actually existing object, namely the man at the gate.

That's the sort of intellectual leap Descartes took. He judged that his concept 'I' was about an actually existing 'I' thing, a *bona fide* real world substance. And the fact that he could suppose this 'I' thing existed in full independence of his material body helped him to suppose further that it was *not* material like his body.

How, one might ask, did he manage to jump so precipitously from word to real world object?—he who scant textual moments earlier had assured us that he *"ought to reject as absolutely false all opinions in regard to which I could suppose the least ground for doubt."[1]* After all, the existence of a word doesn't absolutely certify the existence of a real world counterpart. For instance, when I tell friends about the Leprechaun who frolics in my front meadow every morning, they don't seem to believe that the particular creature I'm telling them about actually exists. And especially in the case of Descartes' proposed 'I' *thing* that is to have *no* dependence on its body, a doubter could so readily respond: "Mr. Descartes sir, even if you've been fortunate enough in life never to have been knocked out cold, haven't you at least met folks who have been? And don't their accounts of the 'nothing' they experienced during such episodes have something pertinent to say concerning your whimsical notion of a *thinking thing* that'll keep on doing its job, no matter what happens to a person's brain?"

THE "THINKING THING"

It certainly is chastening to see how easily even a great thinker like Descartes can fall into error scant moments after he's vowed methodical caution in his approach. I like to think, however, that if Descartes had lived within our current scientifically oriented culture, he'd have been satisfied to render his musings in the form of a *hypothesis* to be explored—not only in regard to its logical cohesiveness, but also in regard to its consistency with empirical findings.

Not, by the way, that current theorists can avoid the notion of a thinking *thing*—given the fact that they have the same cognitive apparatus as Descartes and all the other members of our species. To wit, they have no choice but to conceptualize in terms of objects-and-their-movements. But the current object of choice to underwrite thinking is the brain itself rather than a non material 'I' thing interacting with the brain at some convenient locale like the one Descartes suggested, namely the pineal gland (a singular and centrally located brain structure within an organ whose components otherwise come primarily in pairs, left and right sided).

HOW WAS DESCARTES' NON MATERIAL *THINKING THING* TO INTERACT WITH ITS BODY?

Here's how Descartes' described his putative mind substance:

> ...I thence concluded that I was a substance whose whole essence or nature consists only in thinking, and which, that it may exist, has need of no place, nor is dependent on any material things; so that "I," that is to say, the mind by which I am what I am, is wholly distinct from the body...and is such that, although the latter were not, it would still continue to be all that it is.[2]

So Descartes envisioned a *non* material substance animating his material body. Note that this mind substance was remarkably different from the earlier 'spirit' substance of folklore, which shared with ordinary matter the property of taking up space. His mind stuff was not extended in space *("has no need of place"),* nor was it *"<u>dependent</u> on material things."* Its essential characteristic was simply to be conscious (Descartes used *'think'* in the context of being conscious).

This type of substance greatly increased the difficulties of explaining how there was to be *any* causal interaction with the body. In fact, one might have argued that the total lack of dependence of the putative mind substance on its body implied its inability to be acted on by its body; for if the body were able to send a constant barrage of messages to its mind substance, then the body would be contributing impressively to this thinking substance's states. That is to say, the mind would be depending quite dramatically on its body.

We have to remember, however, that René Descartes was the Michael Jordan of intellect in his day. He was so admired in the sport of philosophizing that his ideas were not likely to be peremptorily dismissed, even though he'd come up with a conceptual model that did not successfully cohere. More pejoratively stated, his model was incoherent; for how was a thing that had nothing in common with another thing to successfully engage the other thing interactively?

Given Descartes' reputation though, some of his readers went to extraordinary lengths to get around this non material substance's inability to interact with its material body. The philosopher/theologian, Malebranche, for instance, proffered the following explanation: God, knowing all things from all eternity, knows exactly what you and I are planning to do next. So God uses each human choice as an *occasion* to cause the body to move as a person has willed it to.

This theory, known as *'Occasionalism'*, dealt with the intractable problem then by kicking the causal factor directly back to the 'First Cause'. Since God is All Knowing and All Powerful—Malebranche reminded his Christian audience of that time—God could keep track of all these human choices without the slightest difficulty, and He could directly execute the needed causal activities with no problem.

Leibniz (he of the best of all possible worlds) made it even easier for God by recognizing that 'He' had the power to set this process up from the beginning via a program to be labeled *'Preordained Harmony'*, a program that God was to have bundled right in with His original act of creation. Since, that is, God knew from All Eternity what we're going to will at any given moment, He could've also set up an eternal harmony between our human choices and our bodily movements.

If God actually did that, however, He sure fouled up when He set the software package in place to cover my own particular instance, because I can't tell you how many times I've *willed* my body to do stuff—like hit a decent tennis shot down the line—only to have my body behave...*uh*...inharmoniously. But no matter, Scholastics of that earlier day also found such explanations much too strained to take seriously, even though these accounts did provide logically coherent explanations within the context of a Personal God whose Act of Creation focused centrally on our own grand and glorious species.

For his part, the young Spinoza delivered a rather biting criticism of his older colleague's thesis. I think Spinoza's unaccustomed asperity on this occasion might've come from his frustration at the gap between the wondrously high standard of truth that Descartes had espoused and his almost immediate lapse into humbug. At any rate, Spinoza twitted Descartes pretty mercilessly about the alleged interaction that was to occur between the non material mind and the brain's pineal gland:

> "I am lost in wonder that a philosopher who had stoutly asserted that he would draw no conclusions which do not follow from self-evident premises...could maintain a hypothesis, beside which occult qualities are commonplace...I should much like to know what degree of motion the mind can impart to this pineal gland, and with what force can it hold it suspended? For I am in ignorance whether this gland can be agitated more slowly or more quickly by the mind than by the animal spirits [i.e. the natural brain mechanisms]..."[3]

SELF CONSCIOUSNESS

On the plus side for Descartes, his conclusion was congruent with the ongoing opinion of Scholastic philosophers, if only in one crucial respect. They also believed that we have a *non material* mind/soul. And one of their favorite arguments *also* rested upon the fact of human self consciousness, a capacity which they believed we had a total lock on. That is to say, although (unlike Descartes) they attributed consciousness to lesser animals, they judged that none of the others was explicitly conscious of itself. Let me illustrate now this wondrous capacity: I can look out my study window at the moment and become directly conscious of the twin pine by my front gate. Then I can turn back completely on myself, so to speak, becoming explicitly aware that it is I who has become conscious of the tree. In sharp contrast, although Shep can become conscious of the tree (especially if a squirrel has just scrambled up its roughened bark), he gives no evidence (Chapter 1) of having the ability to become explicitly aware that it is *he* who has just perceived the tree.

For humans to perform the mental contortion just described, Scholastics judged that we need to be able to turn back <u>completely</u> on ourselves, an act they considered undoable by a merely material organism. Their argument amounted to a thought experiment that went like this: Take any physical object, as pliable as one might wish, say, a thin piece of paper, and try to turn it back *completely* on itself. You might be able to accomplish an extremely tight curlicue, but you'd never get it to turn totally in this fashion. Why? Its parts, flexible as they may be, are extended in space, so at some point, some parts of the item will get in the way of other parts while they're in the process of trying to curl back completely. But since, in magnificent contrast, a non material object has no parts extended in space, there's nothing to get in the way of its turning totally on itself. Hence, a non material (spiritual) substance is both capable and *necessary* in order to perform acts of self consciousness.

I don't know whether readers will be impressed by this sort of proof, but I would gather that current theoreticians of mind find it pretty underwhelming—a conclusion I've arrived at on the empirical basis that I've never seen the above argument mentioned in modern accounts. In any case, as we saw in Chapter 1, since some other animals give good evidence of self consciousness (chimps at the very least), we'd have to end up gifting such lesser creatures with non material souls also. That's a conclusion Scholastic theologian/philosophers might find a bit awkward. At any rate, we will return later to this humanly important issue of *self consciousness*.

AN IMPORTANT CONSEQUENCE OF DESCARTES' TWO PART DIVISION OF THE WORLD INTO MIND AND MATTER

Despite the failure of Descartes' two-substance theory of body and mind, his thinking probably contributed to the scientific community's growing enthusiasm for studying the brain at work. That's partly due to his conclusion that the other animals amount to no more than clever machines without consciousness. *(That in turn followed from the fact that our species, renowned for its conceptualized thinking processes, was to be granted exclusive possession of the remarkable consciousness-substance.)*

The door was thus opened wide for systematic exploration of the cognitive systems of other animals <u>without any need to insert a</u> *<u>thinking thing</u>* for the purpose of viewing what had been detected by the animal's sensory systems, or in order to process the information gleaned, or in order to subsequently decide what course of action to take. Without a *mind thing,* that is, detection and information-processing systems were left to do the work on their own. Turned out this approach succeeded so well that the notion of a *homunculus* at the control panels became superfluous in the case of other animals—and eventually also in regard to our own grand and glorious species. We'll focus on that issue later when we get to examining the current *Standard Model* of consciousness, but first we need to tie up some loose ends in regard to Descartes' historic model.

23

THE MIND APART FROM ITS BODY

In the last chapter, an imaginary critic posed the question: "Mr. Descartes sir, even if you've been fortunate enough in life never to have been knocked out cold, haven't you at least met folks who have been? And don't their accounts of the 'nothing' they experienced during such episodes have something pertinent to say concerning your whimsical notion of a *thinking thing* that'll keep on doing its job, no matter what happens to a person's brain?"

Not just Descartes, of course, but anyone who's had the good fortune never to have been cracked unconscious must take such 'nothing' reports on the authority of those who *have* suffered knock out blows to the head. Some who've not had the experience try to imagine what it's like by making a comparison with, say, deep sleep. Problem is that lesser levels of consciousness persist during much of our sleep time, so we virtually never wake up with a feeling of having come back from nothingness.

Reports of "near death experiences" have led to additional confusion, because the persistence of consciousness during these ultimate moments (when we might assume our brains have stopped functioning) leads some to believe that human consciousness is indeed separable from its bodily container. So to clarify the situation, let's examine now what's involved in such happenings.

NEAR DEATH EXPERIENCES

Sometimes, so-called "near death experiences" have occurred in conjunction with temporary stoppage of the heart. And since in the old days doctors used cessation of the heart beat as their mainstay for declaring a person dead, successful resuscitation has even been referred to rather dramatically as "coming back from the dead." Those who've made this back and forth voyage sometimes report vivid

154

experiences that fall into familiar patterns. Patients have described bright light at the end of a pathway or tunnel, for instance, often associated with feelings of intense well being. Vivid recollection of past events (life flashing before one's eyes), or words spoken by loved ones or by a "higher power" occur also—as do "out of body" experiences, in which affected individuals look down on themselves and their surroundings (almost always from this bird's eye perspective).

Of these events, "out of body" experiences are most relevant for our present purposes. After all, if people retain vivid awareness of themselves when they're *actually* outside their own bodies, then we'd have realistic reassurance in regard to our personal survival after death. So let's focus for a moment on this particular happening.

'OUT OF BODY' EXPERIENCES

First thing we need to be clear about is that during experiences of the above sort, the brain *is* indeed still alive and functioning, albeit in an intoxicated state brought about by temporarily impaired circulation. Under these pathological conditions, insufficient oxygen is available to harness enough energy for the *normal* operation of nerve cell machinery. Additionally, metabolic breakdown-products of a toxic nature tend to pile up. The result is not a non functioning brain, but rather a *mal* functioning brain.

Secondly, we need to make a distinction between a *'mental event'* and an *'outside world event'*. To illustrate, let me recount one of my recent dream experiences: I found myself at my childhood home on Saint Rose Hill in Jamaica Plain, looking across at the Arnold Arboretum. Now, the fact that in my dream I had moved a bunch of miles from my present home would hardly be accepted as proof that I'd *actually* left my own bed (with or without my body).

Same thing applies to the experiences a person has during other states associated with altered brain function. Although we should emphasize that most of these accounts are in fact truthful (there'd be little reason in most instances for patients to provide erroneous reports), their content attests only to the existence of *'mental events'*, not to *'outside world events'*. Why then do experiencers sometimes claim real world validity for the "out of body" event, when presumably they would not do so in the case of their nightly dreams? In part perhaps because subjects get to view their own bodies instead of the Arnold Arboretum—and also because such "near death experiences" are so utterly vivid (indeed people often attribute strong real world implications to sleep dreams that are exceptionally graphic). Additionally, for those who've just gone through such a harrowing event, the content of their experience (most often reinforced by feelings of intense well being) provides immense comfort when responding to a close brush with perhaps our most deeply rooted fear. Death.

In summary, the fact that some types of disrupted brain function become associated with altered states of consciousness rather than with no consciousness at all offers scant support for belief that human consciousness continues in the absence of a functioning brain.

DESCARTES VS. THE SCHOLASTICS

In our last chapter, we started to compare the two-substance consciousness model of Descartes with the model proposed by Scholastics. And we noted that "his conclusion was congruent with the ongoing opinion of Scholastic philosophers, if only in one crucial respect. They also believed that we have a *non material* mind/soul." The contrast between models is striking nevertheless, and worth our taking a somewhat closer look at, especially since the older Scholastic model offered the potential advantage of more accurately reflecting the degree of *unity* that we perceive in human nature.

HUMAN NATURE AS A UNITY

To begin with then, the Scholastics frowned on Descartes' notion of an entirely separate *thinking substance*. We are not, they argued, a mere composite of two different things, and any attempt to make this sort of radical division violates the obvious unity of our nature. In marked contrast, their own conceptual model—drawn from Aristotle—offered the potential of providing a more together vision.

Here's the background: Aristotle had viewed the objects of our world as being made up of *"prime matter"* and *"substantial form."* While these two phrases might superficially suggest two separate substances, that wasn't the case. Rather, Aristotle was focusing on the common sense fact that all *matter* in our world has one or another *form*. So he wasn't positing two different things, but rather <u>two different aspects of the same thing</u>.

Commonly, Scholastics illustrated the point by means of an artist's sculpture—the chunk of marble standing in for *"prime matter,"* and the chiseled figure representing the *"substantial form."* In our actual world of course, there's no such thing as freestanding 'prime matter'. Existing matter always has one or another form (if only the rough irregularity of freshly quarried marble). And in turn, all forms require some matter to 'inform'. *Alice in Wonderland* notwithstanding, the Cheshire smile isn't going to be hanging around after the Cheshire Cat has gone away.

Some forms are obviously more complex than others. The form of Venus de Milo, for example, is more elaborate than that of a marble cone. Progressing in this vein, Scholastics envisioned some forms as being far more elaborate than any of those informing inanimate objects. Some forms, that is, were to impart the unified functioning of living things. Scholastics referred to such forms as *vital principles* (sometimes translated as 'souls'), with different sorts of 'vital principles' accounting for the three levels of life (vegetative, animal, and rationally self conscious).

Interestingly, the Scholastic explanation of human consciousness, though anciently derived, is much easier to square with current theory than the historically more recent thesis of Descartes. The reason is that modern biologists are in fact heavy into *forms*. That is, they don't think of living organisms as being composed of some special kind of stuff. Rather, they recognize that living organisms are made up of the same kind of matter found in non living things. What's different is the way the matter is configured into closely integrated systems composed of and constructed by specialized alignments of matter. And when we start thinking in terms of *'configurations'* and *'alignments'*, we're in the same ballpark provided by the more abstract word *'form'*.

So much is this the case that one might wonder why modern theorists of mind almost always start with Descartes and almost always ignore preexisting Scholastic theory—especially since Scholastic thought almost certainly exerted a strong influence on Descartes. He might not, for instance, have made such a facile jump to the putative existence of a *non material* thinking substance if he'd not been raised in an intellectual atmosphere where such theoretical stuff was as common as angels dancing on the head of a pin.[1]

THREE REASONS FOR IGNORING THE SCHOLASTIC MODEL OF MIND

I think there are three probable reasons why current theorists have ignored the Scholastic (*pre* Cartesian) theory of consciousness; and since these reasons involve factors that are highly pertinent in the construction of <u>any</u> helpful conceptual model, it'll be worth our while to take a look at all three considerations: **1.** familiarity with the basic building blocks to be used in a model's construction, **2.** ease with which a model's components facilitate empirical exploration, and **3.** freedom from arbitrary constraints imposed by rigidly preconceived notions.

FAMILIARITY WITH A MODEL'S BASIC BUILDING BLOCKS

Let's start with a happy example. In Chapter 3, we noted that *when fashioning a conceptual model of the eye, it was natural to think of a camera. After all, cameras also "picture things," and since these devices had been engineered by us humans, we knew what we were talking about.* Implied in this assessment was the fact that all educated people nowadays are familiar with the rudiments of how a camera works, so they're already up to speed in regard to the basic building blocks used in fashioning a camera model of eye function.

In similar vein, the Aristotelian conceptual model of matter and form was familiar to all properly educated people at the time of Descartes (in fact, personal familiarity with the writings of Aristotle was a defining element in being "properly educated"). Hence, the Scholastic notion of *forms* underwriting the activity of living things started with basic conceptual building blocks that were at least somewhat familiar to most readers of the time.

I suspect, in sharp contrast, that some of my present readers have never even heard of Aristotle's theory of *prime matter and substantial form* (at least until the noetic moment provided above). And the same lack of familiarity undoubtedly holds also for some current theorists of mind. That's because, with the exception of antiquarian philosophers, not too many folks read Aristotle nowadays—and those who do are usually struck primarily by the quaintness of his expression (except perhaps in the case of his *Nicomachean Ethics,* which can still touch people where they live).

The upshot is, most *modern* theorists of mind would have to familiarize themselves with Aristotle's basic building blocks before they'd be able to make too much sense out of the Scholastic model of mind, and given the two negatives that we're about to get to, theorists probably think they've got better things to do with their time.

FACILITATING EMPIRICAL EXPLORATION

We noted earlier that "modern biologists are in fact heavy into *forms.*" Not that they're likely to use that particular word, though they often enough use similar terms, for instance, *configurations* and *alignments.* The latter words seem less abstract, however, and more likely to evoke visual images of *structural conforma-tions* apt for performing particular functions.

Let's illustrate now how this more palpable level of cogitating propelled scientists into investigative mode: They figured there must be an information storehouse within a cell enabling it to perform highly complex functions, including that of self replication. Further, much of the information had to be contained in an animal cell's nucleus, because removal of this vital center brought replication to a halt. The nucleus contained a bunch of material called simply…*uh*…DeoxyriboNucleic Acid. Hence, they speculated that this **DNA** stuff might turn out to be the vital information carrier they were searching for, especially since a cell went to great trouble organizing this material into discrete transmissible units (the *chromosomes)* when it divided into two daughter cells.[2]

That knowledge in turn suggested a program that cried out for exploration: How exactly were the components of this DNA stuff configured? And could it be that the configuration amounted to a way of encoding the necessary information? Turns out of course there was a beautifully simple conformation—the now famous double helix—which stabilizes *sequences* of four small molecules for encoding directions that cells need to follow in order to form their crucial constituents. And note how readily the word 'form' jumps right into place as we talk about organisms. That's because what we refer to as 'life' does not come about merely on the basis of certain chemicals that make up an organism's *matter*. What's crucial is the *form* these building blocks take in order to elaborate the structural and functional components required.

In sharp contrast to this modern mode of thinking, however, when Scholastics used 'form' as one of their conceptual building blocks, they maintained the notion at such a lofty level of abstraction that its meaning remained vague—almost to the point of emptiness. What happened in effect was that the Scholastics then turned the *'form'* of organisms into no more than a vacuous *'vital principle'*, an inscrutable something that did little to explain life and did lots to discourage empirical investigation into the concrete alignments that together constitute the complex web of feedback processes we refer to as 'life'. Naturally, when scientists eventually did gather more detailed information about the specific forms underwriting living processes at a molecular level, the occult theory of *'Vitalism'* simply went by the board.

Bottom line: The Scholastic model of life and mind turned out to be a useless dead-ender of a theory because it was composed of totally vague abstractions that did little to advance intellectual exploration. In fact, given the theory's tendency to spin its wheels (all the time promoting occult properties) it actually discouraged intellectual investigation at a practical level. Even so, we might note that if it'd been set within a modern scientific context, the thesis would've had far more potential than Descartes' old model—the one theorists seem to use quite regularly as the launching pad for their own renderings.

FREEDOM FROM ARBITRARY CONSTRAINTS

The third reason why current theorists have probably ignored the Scholastic theory of mind involves a factor of central importance when it comes to the progress of human thought; namely, while all conceptual models require *some* constraints (mechanical engineers, for instance, regularly include restrictions in their models imposed by notions like friction and inertia), nevertheless there's a time to continue using an accepted theoretical notion, and a time to alter or discard it.

Let's start with a case in which modern scientists demonstrated their ability to dispense with an outdated notion: Physicists had initially assumed that light transmission from sun to earth involved some sort of medium through which the waves of light would propagate themselves (Chapter 9). This assumption made initial good sense since all the other waves we'd come across (e.g. water waves, sound waves) require vibrating mediums. So theorists hypothesized a gossamer substance, which they dubbed the *'ether'*, through which light waves were to move. Problem was, no available evidence supported existence of the ether, and accumulating evidence turned out to be incompatible with its presence. Eventually, therefore, scientists dropped this theoretical entity. Lesson? Change of 'paradigm' is possible in settings where theoretical constructs don't get ossified into unwavering dogma.

But now let's take a look at what happened in the case of the Scholastic model of mind. Based on Aristotle's theory of matter and form, you may recall that *"the Cheshire smile isn't going to be hanging around after the Cheshire Cat has gone away."* Congruently then, when a living body dies and decomposes, its 'form' is gone too. For if its form were to continue hanging around, it'd be equivalent to the Cheshire smile remaining after the Cheshire Cat had left town.

But unfortunately that's exactly what Scholastics were *required* to maintain. Why? Because their Religious Institution insisted on a mortal body <u>and</u> an *immortal soul*. But did this not then make of the Scholastic thesis a two-substance theory, and in that respect much like the two-substance theory of Descartes that they'd so roundly criticized? After all, the very definition of a substance involves <u>something that can exist by itself, without having to inhere in something else</u>. So if the human 'form' could continue to exist without the 'matter' of its body, then this form would indeed be a separate substance.

Scholastics had no logical choice but to accept this position. In any case, they'd also enthused over an adjoining proposition that the 'human soul' is naturally immortal because it has no parts—hence, there's nothing to be taken apart, which is how material things get destroyed (Chapter 6). But how then were Scholastics to escape their own criticism of Descartes' two-substance theory, namely that such a composite violated the unity of our nature?

Here in effect was there response: "Well, okay, the non material vital principle is a substance, but you see it's...*uh*...an *incomplete* substance; and it joins with another incomplete substance, its body, in order to form one complete substance." How the Scholastics managed to keep a straight face while rendering this conceptual travesty may be hard to explain. But that's the sort of difficulty you're in for when saddled with a die-hard constraint imposed by rigidly preconceived notions. Scant wonder under the circumstances, however, that modern theorists had better things to do with their time than to take this theory too seriously.

AN ASIDE TO CHRISTIAN READERS

Let me end this section with some remarks addressed to Christian readers who might find the above analysis offensive. When Jesus said to the 'good thief' who was suffering on an adjoining cross, *"this day thou shalt be with me in paradise,"3* he decidedly did <u>not</u> say to him "your immortal soul will be with me in paradise." It was later theologians, influenced by both our natural dualism (Chapter 21) and by the eccentric brilliance of theorists like Plato, who came up with the notion of *mortal body & immortal soul.* Problem was that once this thesis got grooved in, it became rigidified into religious *dogma*—in the most formidable sense of the term, because it was purported to have been revealed by God Himself. And once *that* notion took hold, no orthodox theorist could reject the hypothesis.

But if mere humans in the form of TV writers for *Star Trek* could come up with an alternative theoretical device, then certainly a Transcendent God would be up to that task Himself. I'm referring to a transportation mechanism housed on the Spaceship that could read out the type and position of each and every atom in a human body, then rapidly transmit the information to a distant destination where people could be reconstituted—atom for atom in perfect realignment—just as they were on the Spaceship before setting out on their journey.

It's doubtful that we humans will ever have the power to actually implement a machine requiring such exhaustive detection and computation abilities. But if one starts with the concept of an Omniscient and All Powerful God who is intimately involved in human affairs, the project would be a piece of cake for Him. So New Testament requirements could easily be achieved without having to fall back on the logically and empirically blemished notion of an indestructible soul substance that's totally separable from its body.

LOOKING AHEAD

We've seen that Descartes' mind/body dualism just didn't work. Also, that although the Aristotelian/Scholastic theory of *matter and form* had greater explanatory potential, it became ultimately mired in vagueness and infected by a similar substance dualism. In reaction to the failure of these theories, current conceptual models have simply dropped the notion of two different substances. Hence, such theories are referred to as <u>monist</u> because they involve a single substrate instead of two different ones. Let's turn our attention now in that modern direction.

24

MODERN MODELS OF MIND

It's standard procedure for biologists to correlate structure and function. For instance, our musculoskeletal system is built of *bones* that are movable as rigid units around flexible *joints* in response to the activation of contractile cells grouped into *muscles* that are attached at both ends by *tendons* to adjoining bones. What occurs as a result of the properly coordinated activity of all these components is bodily motion. Abstract theorists would express the point by saying that the property of *bodily motion* <u>supervenes</u> on the occurrence of these factors acting in concert at more basic levels.

Or take the gastrointestinal tract. It's constructed of organs that grind food and mix it sequentially with various enzymes, resulting in the breakdown of complex chemicals into simpler ones that are suitable for use by our own bodies. We wouldn't refer to any of the individual components as digestion, but we could validly say that digestion comes about as a result of their coordinated endeavor. Once again, theorists would express the point by saying that *digestion* <u>supervenes</u> on these more elementary activities.

And we don't even need living systems in order to introduce the notion of *supervenience*. For instance, we can set a pot of water on a stove where the heat energy transmitted to the water molecules will eventually get them jumping around. At the macroscopic level of ordinary vision, what we get to see at this point is bubbling water with steam vaporizing on its surface. Theorists would express what's happened by saying that *boiling* <u>supervenes</u> on the energetic movement of water molecules occurring at a microscopic level.

THE NERVOUS SYSTEM

If we move now to the nervous system, we can also make a connection between structure and function. Networks of highly responsive cells with connecting cables have the ability to register and transfer messages rapidly, all the while performing numerous computations over selected pieces of data that've been brought together from multiple sources. As above, we can express the point at a more abstract level by saying that *information processing* supervenes on the correlated activities of these more basic elements. And the reason I've been underlining this notion is, as we shall see later in this chapter, supervenience is central to the current *Standard Model* of consciousness.

THE *MIND* OF A DEERFLY

But first, let's revisit the deerfly (Chapter 1) in order to illustrate the sort of information-processing activity that's key to its adaptive behavior. To start with, the deerfly's nervous system has sensor cells for sampling relevant energy from its surroundings. On the system's other end, motor cells provide directives to its muscle groups for life promoting activity. In between are a series of computing cells, whose job is to make sense of the different inputs and to initiate motor outputs appropriate to the situation at hand.

In a given instance, the deerfly's visual *sensor cells* may detect a constellation of ambient energy that its *computing cells* read out as vertical-end-stopped-object in motion. This information is then shared with *motor cells* whose job it is to provide approach instructions to various muscle units. Seconds later, visual sensor cells furnish new information that computing cells read out as near-enough-to-commence-landing, and initiate a motor program to achieve this goal. Moments later, when touch sensors become stimulated, computing cells read this sequence as "mission accomplished," and send messages that activate programs for feeding behavior. Under optimal circumstances the deerfly now gets to have lunch, with a white-tail deer picking up the check.

Also at about this point, computational programs within readers may be starting to elaborate dissonance messages between the content of the above two paragraphs and the section head. For what's all the described activity got to do with a deerfly's 'mind'? And who in fact would be silly enough in the first place to attribute anything like a *mind* to this primitive critter?

THE CONCEPT OF MIND

Before responding to that specific question, we might review the notion discussed in Chapter 12 that words are like boxes, and often *"It's as if we've placed an order that should come in a standard-size box, but the box turns out to be so variable in size that more items or fewer items get packed into recurrent orders. And worse still, not only is the box likely to be of different size the next time we order it up, some items we thought we were ordering might not even get packed inside the next time around."*

'Mind' is of course one of those verbal boxes. Not unexpectedly then, it comes packed on different occasions with contents that vary. My *American Heritage Dictionary,* for example, devotes an entire column of eye-straining print to its definition, yet does no more than scratch the surface of ways in which the word 'mind' has been used.

We saw later in Chapter 12 that professionals provide technical definitions to *"sharpen the borders of the words they use in order to fashion these objects into more dependable and less ambiguous containers."* With that goal…uh…in mind, let's now try to fashion a provisional definition of the term. How about: <u>Mind refers to the cluster of functions that enables an individual to detect relevant aspects of a given situation and then to elaborate appropriate responses</u>.

That definition would undoubtedly be far too broad to suit most people. For instance, automatic door openers at department stores detect stuff relevant to their work, and then come up with an appropriate sequence of responses. Specifically, they detect approaching objects, and respond by opening and later closing the door. But almost no one would want to maintain that an automatic door opener has a mind. Why?

One immediate reaction might be to say that only *animals* have the potential of having a mind—though such a response would be pretty chauvinistic, given the fact that theorists can so readily imagine other entities with minds. Centuries back for instance, Scholastics were reflecting on non material beings, the angels, who were judged to have minds vastly superior to our own (that's what made the bad ones like Beelzebub so dangerous). Modern theorists of course think more in the direction of *highly complex* devices to be constructed at some point via the discipline of Artificial Intelligence.

The latter train of thought raises the issue of whether 'mind' does indeed refer to a *"cluster of functions that enables an individual to detect relevant aspects of a given situation and then to elaborate appropriate responses,"* but that the systems involved must be way more complex than an automatic door-opener. In defense of the deerfly, by the way, we should note that its systems for detecting, computing, and elaborating adaptive responses are many many times more complicated than those of the door-opener. We have described only a few of its mechanisms, since our purpose was simply to illustrate a number of basic points.

But rather than tediously outlining some of the deerfly's additional programs, we can move on to more complex animals. Do my sheep, for instance, have minds? And if one thinks it's more than mere metaphor to speak of "the mind of a sheep," then perhaps it *is* the <u>degree of complexity</u> that's lacking in more primitive organisms.

In any case, I hope what's coming across is that people disagree about when to appropriately apply the term 'mind'. And one reason for this lack of consensus is that the word has its primary application within our own species. Disagreement occurs then whenever we begin to extend use of the term to other species—and indeed more than a few theorists would insist that *no* other species possesses a mind, at least in any literal sense of the term. Perhaps our best course then, in order to get clear on what's involved, is to proceed directly now to our own grand and glorious species.

THE HUMAN MIND

Recall that our human cognitive apparatus is hard-wired in such a way that we can deal *only* with objects-and-their-movements (Chapter 11). In this regard, our conceptual system has echoed the way our perceptual apparatus works—a mechanism that was already up and running for millions of years before our human thought system came along. But given the likeness in mode of operation, there's a strong tendency for us to think of our conceptual objects as similar to the sorts of objects provided by our perceptual systems. Specifically, when we think about THE mind, it's easy to conceive of it as an *entity* that performs specific tasks like thinking—much as an automobile engine is an *entity* that performs specific tasks like powering my pickup's wheels.

We saw that this 'thingifying' operation is one of the basic reasons for our dualism at a practical level (Chapter 21), and it helps to explain also why Descartes jumped so readily from the fact that he was thinking to a *'thinking thing'* (Chapter 22). And since both Descartes' *thinking thing* and the Scholastics' *non material vital principle* amount to spirit substances breathed exclusively into human nature, we're given a reason then to conclude that only we humans have minds—since only humans have this mind substance.

But once we've found substance dualism inadequate, the mind ceases to be a *thing*. Rather, we have to look at the word 'mind' as a shorthand term for encom-

passing various activities we think of as mental—processes that inevitably originate in a functioning brain.

Let's go back now to the *American Heritage Dictionary* and look at the first
part of its initial definition: "*1. (in a human or other <u>conscious</u> being) the <u>element</u>,*
<u>part</u>, <u>substance</u>, or <u>process</u> that reasons, thinks, feels, wills, perceives, judges, etc..."
Note that if we were substance dualists, we could choose *"substance,"* and if we're
substance monists, we can choose *"process"*—or the *"element"* or *"part"* (the brain)
that underwrites the process. But even before this dictionary definition gets to
providing us with a sample of mental activity, it starts right off with the notion of
a *"<u>conscious</u> being."*

THE CONSCIOUS MIND

And indeed one of the central features we tend to take for granted in our use of
the word 'mind' is that minds are conscious. In fact, before Freud and others
made their intellectual contributions around the turn of the *last* Century, mention of "conscious mind" would've seemed redundant. Now of course we're
accustomed to the mention of mental processes that are *not* conscious (the sort of
thing illustrated by the devout nun in Chapter 18), and such discussions sometimes use the expression "unconscious mind." The notion of unconscious mind,
however, almost always arises in the context of the conscious mind. For instance,
Freud was big on helping people make repressed (unconscious) material available
to their conscious minds.

So returning to the deerfly, we might ask: Is this insect conscious of its activities? Because if it isn't—if there's "nobody home," so to speak—then very few
would be inclined to speak literally of "the mind of a deerfly." And the same goes
for my sheep.

Not surprisingly then, the issues of when to impute consciousness and when
to make literal use of the word 'mind' go closely together. And in both cases, the
question is subject to ongoing disagreement. One rule of thumb people seem to
use intuitively is that the more similar an organism is to ourselves, the more likely
we are to believe that it has conscious experience.

Let me illustrate: One person I polled recently thought insects were not conscious; also that frogs and snakes were probably not conscious. She thought that
birds were most likely conscious, and that sheep indeed had this amazing
attribute. "How do you go about deciding?" I asked. Her immediate answer: "I
guess if I can relate to them."

People I've polled who've been exposed to modern philosophy have sometimes answered my question with a question of their own: "What do you mean
by consciousness?" In response to this request for clarification, I've made my
query more concrete by reference to one of our all too common types of experience: "Well, if you injured it, would it feel any pain?" Most of the people I've
talked with decide then that mammals are conscious, whereas many don't think
insects like deerflies and mosquitoes are.

MONIST THEORIES OF CONSCIOUSNESS

Current theorists of mind have the same sort of uncertainty as ordinary folk when
it comes to deciding when consciousness makes its appearance in the animals on
our planet. In pursuing the issue, they usually start with the one instance of consciousness that (almost) everyone agrees on, namely our own grand and glorious
species. And evidence from members of *Homo Sapiens* indicates: if no functioning brain, then no consciousness; if normally functioning brain, then consciousness.

Theorists next ask: What interacting systems within the human brain are necessary for consciousness to occur? It would follow of course that if a particular
type of animal doesn't have the complex judged necessary for consciousness to
make its appearance, then consciousness would not be attributed to such animals.

Sometimes, theorists try to infer the sorts of elements that would be necessary
in *any* functional system in order to produce consciousness. That's because, in
principle, if one could fashion such a system out of, say, silicon chips instead of
carbon based nerve cells, then *that* apparatus would also show the property of
consciousness.

THE *SUPERVENIENCE* OF CONSCIOUSNESS

Back now to the word we introduced at the beginning of this Chapter. We saw
that boiling *supervenes* on the microscopic movement of highly energized water
molecules; also that bodily motion *supervenes* on the interaction of basic elements
in the musculoskeletal system, that digestion *supervenes* on lower-level operations
within the gastrointestinal tract, and that information-processing *supervenes* on
the activities of nervous-system constituents.

In similar fashion then, theorists want to say that consciousness *supervenes* on
certain types of activity within highly complex nervous systems. And since that

notion lies at the core of our current *Standard Model* of consciousness, let's turn next in this direction.

25

THE SUPERVENIENCE OF CONSCIOUSNESS

We saw earlier that Scholastics ran into trouble with their theory of *matter and form* because their concept of 'form' as applied to living organisms remained so utterly vague. This happened of course when their armchair imaginations hit a brick wall. For it was one thing to envision the relatively simple configurations of a Grecian urn or a Roman statue, but the Schoolmen couldn't even begin to imagine the many tiers of molecular complexity needed to supply functional form to even the simplest of organisms. So they ended up turning the complexity-of-form possessed by living things into a *mysterious something* they referred to as a 'vital principle'.

It's important for us to keep this lesson in mind, because if we allow our use of 'supervenience' to drift toward a similar degree of vagueness, we could find ourselves adding yet one more *mysterious something* to an already tiresome list of occult properties that have burdened thinkers throughout human history. In order to avoid that sort of problem, let's nail the notion securely now to our everyday world.

ANCHORING THE NOTION OF 'SUPERVENIENCE'

When we referred to water boiling as a phenomenon that *supervenes* on the highly enhanced motion of its constituent molecules, we might have said instead: the boiling of water *amounts to* the sum of the movements of all its molecules after they've become highly energized. No mystery there! And when we referred to bodily motion as a phenomenon that supervenes on the activation of muscles attached at both ends to adjoining bones, we could have said that bodily motion results from the coordinated interaction of these components. Again, no mystery!

170

And we could also explain, in admittedly more complex fashion, how information-processing *supervenes* on sequences of change within cells of the nervous system. Recall, for instance, the sample of computational activity we reviewed earlier (Chapter 3) involving the visual system: Detector cells in the retina respond to certain wavelengths of electromagnetic radiation and forward messages about this stimulation to 'on center' ganglion cells. Detector cells within the 'bullseye' of a detector-grouping that reports to a given ganglion cell will stimulate the ganglion cell in the direction of firing off a message, whereas detectors located outside the bullseye will tend to inhibit the ganglion cell's firing. The amount of inhibition received by the ganglion cell gets subtracted, so to speak, from the amount of stimulation it receives, and if the remainder still provides sufficient energy, the ganglion cell will fire off a message to the brain. This *amounts to* a tiny instance of computational activity that gets played out simultaneously by a host of additional ganglion cells at the same lower level within the system.

As we also saw, computations are repeated in level after level of connected nerve-cell networks. So, for instance, a particular 'edge cell' in the visual cortex will eventually receive messages from the above ganglion cell and from other ganglion cells running at a given angle from the first. If the sum of stimulation that arrives from this 'straight-line grouping' of ganglion cells is sufficient, the edge cell involved will fire off messages to other brain cells. Layer after layer of such calculations occur—activity that spreads also to non visual networks, resulting in the sort of computational cascade we refer to as *information-processing*. A bit more complex than the above instances, but once again, no mystery!

APPLYING THE NOTION OF *SUPERVENIENCE* TO CONSCIOUSNESS

As information-processing supervenes on functioning of the nervous system, so does consciousness—at least in the respect we mentioned earlier. That is, If there's a normally functioning human brain, consciousness will occur; and if brain function is severely compromised, consciousness will not occur. More specifically, current theorists of mind connect consciousness with certain types of information processing; as, for instance, when information becomes broadcast widely throughout the brain, or when certain types of 'higher level' systems register the fact that lower levels of information-processing have occurred within constituent subsystems.

Yet a thoughtful person might still wonder: How does *any* spectrum of information-processing, complex as it may be, amount to or give rise to consciousness?

For the linkage hardly falls into place with the obviousness of instances like those just described. If that consideration concerns present readers, they might follow the advice given by theorist Daniel Dennett:

> It turns out that the way to imagine [how all that complicated slew of activity in the brain amounts to conscious experience] is to think of the brain as a computer of sorts…By thinking of our brains as information-processing systems, we can gradually dispel the fog and pick our way across the great divide, discovering how it might be that our brains produce all the phenomena.[1]

Problem is that despite such encouragement, most students of the subject have been unable to pick their way happily across this great divide between the occurrence of objective computational processes and the subjective experience of consciousness. That is to say, how information-processing of *any* ilk would amount to consciousness doesn't seem evident to them.

THE SUPERVENIENCE OF NOVEL PROPERTIES

Hence, some theorists have retrenched in the following fashion. They note that over the span of human inquiry, many instances of novel properties have been observed to supervene on the activity of underlying systems, even though humans were not yet able during a given era to understand exactly what was going on. Take even the simple instance mentioned earlier of water boiling on a hot stove. Whenever there was a 'system' composed of water, pot, and hot stove, our ancient forbears knew as well as we that boiling will occur. But to them the connection was little more than a brute fact of nature. After all, they had no knowledge of H_2O molecules and no kinetic theory of heat as being reducible to the collective motion of such fundamental units.

On a base of analogies like this, current theorists point out that our own understanding of what's happening in the sphere of mind-and-body is still at a relatively primitive stage. That is, we know a normally functioning brain *will* bring about consciousness, although we don't yet know how. But we know enough to know that this linkage represents another of those instances when *if A occurs, then B will also*. In that sense, they argue, it's completely appropriate for us to speak of consciousness as supervening on the highly complex activity of our human brain.

A PROMISSORY NOTE

Within such a context, however, the current *Standard Model* of consciousness leaves us, not with any clear understanding about how brain functioning gives rise to consciousness, but with what amounts to a promissory note that *someday* we will know. Unfortunately, we're left in the meantime with no more than a particularly nebulous use of the verbal container 'supervenience'—a situation that's all too reminiscent of how Scholastics handled the same problem: the Schoolmen of old resorted to a mysterious *'vital principle'*, and some of our current academicians resort to an equally mysterious type of *'supervenience principle'*.

And what's even more daunting is the fact that our promissory note has nothing even close to a triple A rating, so we may never get to collect. That's because, while science of the human brain continues to progress with truly amazing rapidity, the arrow of our increasing knowledge about human consciousness has been pointing *only* in the direction of <u>correlation</u>. That is, specific brain areas/functions are being progressively correlated with particular types of consciousness. But within such ongoing studies, the occurrence of consciousness itself is treated simply as a given. Hence, a question like "How does the brain generate consciousness?" is not even asked. Instead, objectively answerable questions are pursued, such as "What systems in the brain must be up and running in order for any type of consciousness to occur?" and "In order for a person to experience a given type of consciousness (e.g. bodily pain), what systems must be operative?"

CORRELATING CONSCIOUSNESS WITH BRAIN FUNCTION

We noted initially that without an adequately functioning brain, consciousness will not occur; also that when our human brain *is* functioning normally, consciousness will inevitably make its occurrence (a statement that needs a bit of qualification, since there are normal portions of brain activity during deep sleep that are devoid of consciousness). We started with that basic fact because even scientifically primitive humans have had the elementary data available that would allow them to correlate knockout blows with loss-of-consciousness (Descartes notwithstanding). But by the mid nineteenth century, neurologists had progressed to the point where they'd become able to relate even certain types-of-consciousness to the activity of specific brain areas. They were able to achieve these advances by 1. making careful note of functional deficits shown by patients who suffered from, say, strokes, 2. following these patients for the rest of their lives, and 3. then performing careful examinations of the involved brains after the patients died.

Two famous instances are named after the physicians who were able to correlate the loss of ability to become conscious of words with the specific areas of brain that had been destroyed. In *Wernike's aphasia*, destruction of tissue high on the left temporal lobe results in loss of ability to understand the spoken word. Patients hear the sound okay, but they have no consciousness of the word's meaning. In *Broca's aphasia*, destruction of brain tissue well in front of the motor strip on the left frontal lobe results in loss of ability to have words pop out of us as if by magic whenever we want to express ourselves. Patients afflicted with brain destruction in these areas *do* remain conscious, but they've lost that precious arena of human consciousness having to do with our ability to use language.

BRAIN IMAGING

By the latter part of the Twentieth Century, scientists had developed techniques of mind-boggling sophistication that allowed them to look in great detail at what was going on within the human brain—all the while avoiding damage to that magnificent organ. Functional Magnetic Resonance Imaging (fMRI) has turned out to be perhaps the most useful of these methods to date because it provides a highly detailed image of the brain, as well as the ability to detect short-term changes in activation of the brain's many elements.

Use of fMRI has allowed a quantum leap forward in our ability to <u>correlate</u> brain activity with types of conscious experience—in comparison to the century before when neurologists had to wait for the occasional accident of nature (e.g. a stroke), and then had to wait for the years that might intervene before a particular brain became available for examination after the afflicted patient died. And being able to detect exactly what systems become more active under given circumstances provides points of information that were not available when observations were restricted to functional *deficits* after the destruction of a given brain area.

Let's illustrate now how the new methods can be used to explore even subtle states of human consciousness: We know that members of our highly sophisticated social species have the amazing ability to put themselves psychologically in the other person's place. When dealing with adversaries, this skill helps us intuit where they're coming from; and when dealing with friends, it helps us to participate in their joys and to feel their pain...*uh*...so to speak (the tag on indicating that we don't exactly *feel* the pain of their broken leg, but we sure "feel pained" at what's just happened to them).

In order to clarify the brain apparatus underlying such an amazing skill, neuroscientists recently studied pain-related empathy in 16 couples, making due note of an *"assumption that couples are likely to feel empathy for each other."²* Brain activity in the woman of each pair was imaged via fMRI during a series of random shocks applied either to her hand or to her partner's hand. During the procedure, she was signaled ahead of time when the painful shock was to be inflicted on her partner instead of herself. On trials when she herself received the shock, the increased brain activity fell into an already well known and rather complex pattern involving two different though overlapping systems: 1. sensory areas needed to fix the location and intensity of the stimulus, and 2. affective areas that provide the horrific component to our experiences of pain.

On those occasions when a woman's partner received a painful jolt instead of herself, the same affective areas became activated in the woman's brain as when she herself received the jolt, although *not* the specific sensory areas that had been activated when she received the painful stimulus directly. And congruently, at the level of consciousness she experienced *distress at her partner's pain*, though of course *not the pain experience itself.* Investigators were even able to document this correlation in a quantified manner, since it turned out that the degree of activation in the designated affective areas correlated positively with scores on two empathy scales that had been previously administered to the subjects. (Although subjects knew what would be required of them in the experiment, they'd not been told they were involved in an "Empathy for Pain" study).

The most amazing thing to me—as about half of present readers may have already noticed—is that scientists did this study the hard way. That is, since men are well known to be much more empathic than women, experimenters should've used the male partners as subjects to be brain-scanned during the experiment instead of the women.

Back to reality. Experiments of this sort demonstrate that even subtle states of consciousness can now be explored by techniques of modern science, filling in with previously unimaginable detail the relationship between brain function and consciousness. So even if Descartes *had* had reasonable reason to deny the dependence of consciousness on brain activity in his day, such a stance would be totally beyond the pale in our present era, when not only the fact of consciousness but all its various permutations turn out to be anchorable in specifiable brain activity.

Nonetheless, modern studies speak only to the <u>correlation</u> of consciousness with brain function. Ongoing experiments reveal nothing about how or why consciousness arises. And though the current Standard Model hinges on what we've referred to as the 'principle of supervenience', the way that verbal container ends up being used tells us no more than *"Well, if the brain is functioning normally, then consciousness will occur."* To wit, the term 'supervenience', as it's being used, amounts to little more than the invocation of a totally nebulous property that adds virtually nothing to our actual understanding (as in "Morphine produces sleep because it possesses a dormitive principle").

PROBLEM OF CONSCIOUSNESS VIEWED AS HUMANLY UNSOLVABLE

Given the situation just outlined, one theorist, Colin McGinn, came up with a now famous suggestion for how to address this problem; namely, for us to admit that there is *no* solution, humanly speaking. His notion sent shock waves through the family of mind-theorists because it seemed as if a respected thinker was giving in to the irrational urge of "mysterianism." But his suggestion appealed to me, especially since it's so congruent with the general thesis we developed in PART I of the present book. Here are some of the things McGinn has to say:

> Conceiving minds come in different kinds, equipped with varying powers and limitations, biases and blindspots, so that properties (or theories) may be accessible to some minds but not to others. What is closed to the mind of a rat may be open to the mind of a monkey, and what is open to us may be closed to a monkey.[3]

McGinn argued that the brain must have some property that accounts for consciousness, but that our human apparatus is, as he put it, *"cognitively closed"* to understanding what that property might be—much as a rat's cognitive apparatus is closed to grasping the principles involved in the use of differential equations. Note that McGinn was *not* urging miracles upon us to account for the occurrence of this wondrous property. Because a rat cannot comprehend the principles of complex mathematics doesn't mean that such principles are supernatural; it means only that this particular critter's cognitive system can't grasp what's involved. In similar fashion, McGinn argued that our human apparatus isn't up to the task of understanding specifically what's involved in the supervenience of consciousness on brain activity.

And while theorists of mind agonize over the problem—some like McGinn report literally losing sleep about it—most folks live their entire lives happily while scarcely giving the issue a thought. And the interesting point is that fuller knowledge of the mind/body relationship isn't at all necessary for successful adaptation to our *Ordinary World.* Hence, there's no compelling reason this capacity would've been naturally selected.

As philosophers long for absolute truth and photographers long for the ability to judge light intensity accurately (Chapter 2), theorists of mind long for complete understanding of the mind/body connection. By contrast, Natural Selection is interested only in *approximations*—and in only certain types of approximations at that, namely those promoting successful adaptation of an organism to its *Ordinary World.* Some approximations of course are better than others; and with that fact in mind, I'd like to get to my own 'solution' to the problem of consciousness shortly. But first we need to address yet another problem of human consciousness, one we might introduce by asking the question: "What does consciousness actually <u>do</u>?"

26

WHAT DOES CONSCIOUSNESS DO?

Suppose you were seated ringside at a heavyweight bout, and one fighter hit the other with a perfectly timed right to the jaw, dropping his opponent to the canvas like a sack of sand. Now suppose that after the ten count, a person next to you identified himself as a psychic, and told you that he'd been the one who actually caused the knockout. How? Seems that just as the boxer had delivered his blow, the psychic transmitted a zap of psychic energy, and it was this energy that had actually cold-cocked the hapless opponent.

I suppose some of my present readers might want to check the claimed psychic to see if he happened to be carrying a bottle of Sneaky Pete, but in any case, even those who were open to the possibility of such 'psychic energy' would likely reject the fellow's account out of hand. Why? Let's take a moment to state the reasoning explicitly: *When one causal account provides sufficient explanation for something that's happened, any additional account becomes superfluous.* Theorists who use this approach invoke The Law of Parsimony, though sometimes they refer to it as *Occam's Razor*, in honor of the famous 14th Century philosopher who wielded this logical instrument tellingly against many of his Scholastic opponents.

Some say that the label, *Occam's Razor*, comes from the fact that William of Occam used the principle to shave off the sorts of superfluous philosophical embroiderings that were common in his time. I've often wondered though whether the phrase caught on because it also describes a habit pattern that seems so common among philosophers—the tendency to revel in hostile put downs of their opponents. And there's no doubt that some of William's opponents got cut up more than an outclassed fighter in the ring, at least verbally speaking. But in any case, when modern theorists of mind apply the Law of Parsimony in the causal arena, they insist on what's known as *'causal closure'*—referring to the fact that once we have a sufficient causal account of a given activity, we have no need for any shadowy additions.

WHAT DOES CONSCIOUSNESS DO?

All of which brings us back to the present chapter's query: What does consciousness actually <u>do</u>? During most of recorded history, the answer would've seemed obvious: Consciousness allows us to detect what's going on, and then permits us to take appropriate volitional action.

An illustration: If someone steps off the sidewalk to cross the street, but suddenly a car comes careening around the corner, the person will most likely spring back onto the sidewalk. If we were to ask him why, he might say: "I saw this crazy driver coming, so I jumped out of his way." That's the sort of *causal explanation* we normally derive from our conscious experience of events.

But let's rerun the whole sequence now, providing a causal explanation at a different level, that of our brain machinery in action: As the person steps off the sidewalk, receptors at the periphery of his retina detect a fast approaching object. A rapid shunting of information causes immediate shift of his central visual area (fovea) in the direction of this object, and information-processing areas in the cortex identify it as a good-sized vehicle. An avoidance motor-program becomes immediately activated, causing the lead leg to extend and the other leg to partially contract.[1] The thrust-off force created by sudden extension of the lead leg causes the body to jounce back in the direction of the other leg, which has contracted sufficiently that it doesn't trip the person up. The flexed leg automatically extends again after a moment, however, producing an agile move that rebalances the body so the person doesn't lose his footing and fall down.

Which of these causal explanations should we accept as correct, the nervous-system based account or the psychological account? After all, we have to choose one <u>or</u> the other, because if the brain's integrated functioning causes this self-protective behavior, the psychological account becomes *causally superfluous*—unless, that is, one were to fall back on substance dualism.

A Cartesian dualist would be able to invoke Harry Truman's *"Principle Of Where The Buck Stops,"* claiming that his consciousness substance was the true cause of the behavioral response, and that the brain's adaptive apparatus was merely the instrument. Scholastics would put it like this: The consciousness substance is the *efficient cause*, while the brain is the *instrumental cause*—much as Lee Harvey Oswald was the efficient cause of JFK's murder, while the loaded gun he fired was the instrumental cause.

The Cartesian approach has the virtue then of avoiding the either/or problem by invoking a *causal sequence*. Step 1: The body's visual system detects the onrushing car and delivers the information to the consciousness substance. Step 2: The consciousness substance quickly decides it'd better get out of the way of this hurtling vehicle, and sends its decision back to the brain's motor system, telling it to make the body jump back. But attractive as this interactive explanation might seem at the moment—given the fact that it integrates the two different accounts into a unified causal sequence—we saw earlier (Chapters 21 and 22) that the notion of substance dualism is fatally flawed. Hence, modern theorists can't fall back on it here without becoming logically inconsistent.

(SEMI) REFLEXIVE BEHAVIOR

One might argue, however, that the sort of nervous-system account provided above works well enough for more or less automatic (reflex) behavior, but it can't render an adequate explanation for thoughtful behavior. So let's examine that point a bit more closely.

'Reflex' is a term we apply to simple stimulus-response arcs within an animal's nervous system that operate automatically once they're set in motion. In some primitive systems like, say, that of the sea slug, these simple arcs operate with admirable predictability. However, when it comes to highly complex brains possessed of layer on layer of interactive neuronal circuits, even the simplest of stimulus-response arcs is subject to 'outside' influences from within the very same nervous system.

Let's illustrate the point by reexamining the knee jerk, a simple two-neuron reflex. In Chapter 15, I said: *"Reason the knee jerk's so predictable is that it involves a simple built-in arc between an incoming nerve and an outgoing nerve."* But that statement needs qualification, because sometimes the knee-jerk response doesn't seem to want to happen. (Any med student trying to learn the art of physical examination will agree with that). Here's a trick that experienced physicians will often teach at this point: Ask the patient to grasp both hands in front and pull, while looking up at the ceiling; then plunk him right below the knee cap with the rubber hammer. *Voila!* The knee jerk will unveil itself like magic. Why? In part because distracting the patient will interfere with the neural inhibition that's keeping the knee jerk from occurring. In other words, even the simplest of reflexes is subject to complex modulation within sophisticated nervous systems.

THOUGHTFUL BEHAVIOR

Suppose, however, that what's involved is not a more or less automatic response like jumping back to the sidewalk for safety. Let's take the case of, say, a mother on a shopping trip who wants to buy her young child a new pair of slacks. The complexities involved in this case are much greater, and the requirement for immediate action less urgent; hence many more sub systems within her brain will end up contributing substantially to her ultimate decision. For instance, memory modules will be required in order to remind her where the store is. When she gets there, primary sensory areas will swing into play to detect the piles of slacks, and higher-level systems will be needed in order to identify the objects as slacks. Evaluative systems will come importantly into play at this point, contributing a variety of positive and negative *weightings*, and these weightings will foster her final decision about whether or not to buy.

If, for instance, the mother's not particularly wild about any of the items, her esthetic weighting system will vote no. But if there's a very good bargain, her sense-of-value will furnish positive input. If her child is growing fast, considerations of time-and-change will provide input concerning the foolishness of paying top dollar for the best quality of clothes available. Furthermore, if the mother shops often and there's no crucial need for a pair of pants immediately, this information will also be factored into the decision. And so on.

Now while the weighting of all these various elements will be *accompanied by* consciousness, the neural back-and-forth among the various identifying and weighting systems will eventually compute the answer—one that's congruent with the mother's learned values (i.e. weighting programs will have been ongoingly tuned by a lifetime of personal experience). In some cases of course, the deliberative influence of all these weighting factors will be short-circuited by a few initial factors, and we refer to people who characteristically operate in such fashion as "impulsive."

CONSCIOUSNESS AS AN 'EPIPHENOMENON'

Once theorists became familiar with our brain-based computational mechanisms, the problem of causal closure dawned on them—as well as a possible solution to that problem; namely, that the brain apparatus brings about the behavior, while consciousness acts only as a causally impotent *'epiphenomenon'*. I've always had the fantasy, by the way, that theorists figured a six syllable word would make the thesis sound more impressive; hence, it'd be taken more seriously. At any rate, one of my favorite illustrations of the notion was provided by William James. As he put it, *"the shadow runs alongside the pedestrian, but in no way influences his steps."* The shadow is merely epiphenomenal.

James described the concept of epiphenomenalism with great clarity but with little enthusiasm. Problem is that the notion is grossly counterintuitive. That's because we have such a vivid sense of our conscious selves making choices—so much so that to take our ordinary judgment in the matter as nothing but illusion is difficult to accept. Of course, we've already seen that we're sometimes forced to swallow highly counterintuitive notions. For instance, scientists had to reconcile themselves to the fact that electromagnetic waves propagate without any medium to do the waving (Chapter 9), something that seemed remarkably counterintuitive at the time.

WHEN TO ACCEPT THE COUNTERINTUITIVE

It's not then that we're never forced to bite the bullet and accept counterintuitive explanations. But most theorists agree we should stick with our normal human intuitions until empirical findings and logic require us to abandon them. That is to say, accepting a grossly counterintuitive thesis should be the last resort, to be embraced only when all else has failed.

One of the reasons theorists can seem so far out when it comes to accepting counterintuitive explanations is that their job requires them to follow a given line of reasoning through more closely than usual. The above issue illustrates that point, because the ordinary person doesn't generally push the notion of *causal closure* all the way to its logical conclusion. And in retrospect, we might add, it's perhaps surprising that so many theorists of mind were so slow to pick up on the issue. But once they did, the fact that epiphenomenalism 'solved' the causal closure problem brought a number of theorists into its camp.

ELIMINATIVISTS

One final act of desperation on the part of theorists buffeted by the causal closure problem was to deny that consciousness exists. After all, if consciousness doesn't *really* exist, then you don't have to worry about it throwing a monkey wrench into the explanatory account that's based on nervous system activity. Nevertheless, such theorists aren't keen on being thought of as crazy, so they're seldom into denying the existence of the obvious in a straightforward manner. Instead, they're likely to shift perspective on consciousness, referring to this very core of our human life as if it were but some outworn theoretical notion.

One of the earliest theorists to take this tack was behaviorist, John Watson, who said that *"belief in the existence of consciousness goes back to the ancient days of superstition and magic."*[1] He thereby treated our assent to its presence as no more than an article of faith. In like manner, he spoke of the *"major assumption that there is such a thing as consciousness,"*[2] thus downgrading it to conjectural status. There have been many eliminativists since that time, and their arguments have usually involved variations around that same early theme.

Eliminativists have also dealt with the pesky presence of consciousness by asking "believers" to define precisely what they mean by the term—a bit of a problem, since as we noted earlier (Part III, Introduction), consciousness is too fundamental to be broken down into more basic components for formal definition. They then conclude that the best way to overcome all this vagueness is to stop using the word 'consciousness' entirely. Problem is that discarding the word won't make consciousness go away.

But nowadays, in order to clarify what they're talking about, most theorists have settled on the phrase *'phenomenal consciousness'* to refer to our normal day-to-day states of wakefulness—what philosopher John Searle referred to as *"those states of sentience and awareness that typically begin when we awake from a dreamless sleep and continue until we go to sleep again..."*[3]

It's a proposed solution for the origin of these phenomena toward which we will now labor. And there's a bit of good news to report about the explanatory story I'm going to be running by you: It nicely handles the *causal closure* problem without having to revert to a two-substance causal sequence and without requiring us to think of consciousness as causally inert.

27

SPINOZA AND PANPSYCHISM

A number of years ago I hand dug a small pond in an area of meadow near the bottom of my land. That sort of activity takes the place of a Nautilus machine in my life—not to mention the fact that my sheep appreciate having a handy water trough available. Because the pond's banks are shaped into the neat geometry of a square rather than forming the sort of enclosed meanderings that characterize the banks of an ordinary pond, I dubbed the result "Spinoza's Pond" in memory of that great 17th Century poet of philosophy.

Not that he ever thought of himself as a poet. For when in his philosophical writing he followed the rigorous approach made famous by Euclid, Spinoza obviously believed he was applying the same sort of disciplined logic in order to achieve conclusions about 'What Is' that were comparable to the inexorably certain conclusions of that renowned geometrician from ancient Greece—or at least "inexorably certain" in the minds of 17th Century thinkers, who still agreed that the truth of Euclid's axioms was self evident. But at any rate, I've always thought it charming that Spinoza used this poetically symbolic mode of exposition (formal propositions, followed by proofs, and then corollaries) to create in himself and others a sense that his logic led to unassailable truths.

Nowadays experts think differently, since they judge that even a supposedly self-evident truth like "a straight line is the shortest distance between two points" breaks down when applied to the vastness of the universe. Worked pretty darn well for me though when I was measuring proper lengths for the pole barn I eventually built to shelter my sheep. Of course, in that undertaking I was ensconced well within our *Ordinary World*.

But even though a much expanded information base requires us 21st Century dwellers to take Spinoza's geometric approach with a touch of poetic salt, he was nonetheless an awesomely brilliant figure, and what he had to say is still worth our attention. Specifically, he saw all-that-is as the manifestation of one substance, and that this one substance has 'attributes' of both *matter* and *consciousness*. His careful reasoning led him to what in modern parlance would be referred to as a <u>dual aspect theory</u>. That is to say, all the 'modes' of existence we see around us (like you and me and the apple tree) have properties of both matter and consciousness. Or in other words, everything has both a material and a psychic aspect—which of course is what's meant by *pan* (all) *psychism*. Let's take a look now at how Spinoza reasoned things out.

THE CONCEPT OF 'SUBSTANCE'

Just as Descartes was influenced by the intellectual tradition in which he was raised (Chapter 23), so was his younger contemporary, Spinoza, who chose the concept of *'substance'* as a centerpiece for his reflections—not surprising, given the fact that ever since Aristotle's time, the notion had been pivotal. In keeping with this long tradition, we've still been using the term in this 21st Century work (77 times already). And yet I suspect that not too many readers have even bothered to ask themselves what the word means, or why I haven't bothered to provide a more explicit definition. If my hunch is correct, the reason you probably haven't registered your dismay at this state of affairs is that the word's meaning seems so common sensical: A substance is, well, you know what I mean…

    We're going to look at the word more closely now though, and since it's going to take a bit of time to do so, let's use the opportunity first to also review one of the core points of the present work; namely, that **our brain's information-processing apparatus employs engineering shortcuts that limit us to (usually helpful) approximations of the world. Specifically, we perceive our surroundings *only* in the form of <u>physical-objects</u>-and-their-movements (changes), and we conceive things *only* in the analogous form of <u>conceptual objects</u> and their manipulations. But wondrous as this latter capacity is, results become progressively problematical as we roam further and further from immediate dealings with our *Ordinary World*.**

ARISTOTLE AND 'SUBSTANCE'

Aristotle started out applying the term 'substance' to the ordinary objects that we see around us. Take my canoe, for instance. It's a substance. Why bother using a word like that when referring to the various objects in our ordinary world? Because it was helpful for certain purposes. So much so that Aristotle kept expanding the uses he made of the term[1]—to the point, it must be said, where one use would start to conflict with another use. If we look at the situation with a bit of humor though, it was the great man's very inconsistency that gave his many commentators an opportunity for refined intellectual fun by disputing among themselves his real meaning. Of course, great as Aristotle was, it's unlikely that he remained totally free from the inconsistency that plagues the rest of us mere mortals. At any rate, for the moment, we'll concentrate on one specific way the term 'substance' came in handy.

If someone unfamiliar with the word 'canoe' asked me what it meant, I could say it referred to a relatively thin vessel that tapers to a point at both ends and that's usually paddled through the water. My own canoe, I might add, has a bright yellow color, except where it's gotten scraped along the bottom from my relentless pursuit of Class 1/2 rapids along the tortuous course of the Charles River, its creek-like status 25 miles west of Boston belying its majestic entrance into the city along the Charles River Basin.

SUBSTANCE AS CONTAINER

At any rate, the *substance* of my canoe acts as a handy container for its properties. Thus, the yellow color and the scraped bottom aren't just waving free in the breeze, so to speak. They're found together, "inhering" within the very same repository (along with a bunch of other properties like the tapering-at-both-ends). And when I use the *word* 'substance', it acts in analogous fashion at the conceptual level, this time as a verbal container in which to store the linguistically described properties of 'yellow', 'scraped', and so on.

Notice that what Aristotle was doing here involved simply the use of our information-processing apparatus in accordance with its *obligatory* manner of operation. He started, that is, by dividing stuff into objects (substances) and their sometimes changing manifestations (movements). He then made use of the relevant conceptual objects and the appropriate properties assigned to these verbal containers—or as philosophers are wont to say, the properties that are to be *predicated of* these containers, as in "My canoe *is* yellow."

OBJECT CONSTANCY

We noted in Chapter 3 that "object constancy" helps us to deal with our surroundings more consistently, and hence more successfully. Aristotle's division of stuff into *substances* and their *properties* amounted to a useful way then of conceptually reinforcing this procedure. For instance, suppose I decide to paint my canoe red. It will now have a different color, but it will still be my canoe—same substance, that is, with one of its incidental *('accidental')* properties modified.

This approach works well as long as we stick to everyday dealings with our surroundings. But what happens if we systematically peal off each and every property that might be predicated of my canoe? It ~~is yellow; it is scratched; it is tapered at both ends~~, and so on. When all the subtracting's done, what in the world remains of the canoe's substance? Nothing at all! At least, nothing we can mention—for whatever we might say would be yet another property that we could then subtract also from the canoe's basic substance. This situation provided much consternation for fancy thinkers. In fact, John Locke finally threw up his hands and said: *"our idea of substance…is but <u>a supposed I know not what</u> to support those ideas we call accidents"²* ('accident' being a philosophical term applying to incidental properties like my canoe's color).

Why didn't fancy thinkers simply dispense with this I-know-not-what? Because then they'd have had to deal with a concatenation of yellow, and scraped, and tapered-at-the-ends, floating around in thin air with nothing to account for why they all stayed so nicely together. And even if theorists had tried this approach—a few attempted to do so—they would've only ended up using some other verbal containers (like the word 'concatenation' that we've just now employed). And qualities like 'yellow' and 'scraped', if we try to leave them in free flotation, end up as objects in themselves anyway, unless we attach them to some "more substantial" object. All this comes about because our perceptual and conceptual systems obligatorily follow an engineering shortcut employing *objects-and-their-movements*. We have no choice, because that's the sort of bedrock program employed by our conceptual systems (Chapter 6).

IMMANUEL KANT'S *"THING* AS SUCH"

When the great Immanuel Kant came along, standing on the shoulders of giants like David Hume and John Locke, it occurred to him that we process the information gleaned by our senses in obligatory ways that go with the territory of being human. For instance, if I walk out my door now, and the Mourning Doves in my front meadow take flight, I will *see* only the one event followed by the other; yet I will automatically interpret the sequence in causal fashion, since that's the way my information-processing apparatus works (Chapter 8). That is, I will 'see' my approach as *causing* the birds to fly away. Kant reflected on such occurrences at a time when little was known of neuroscience and well before the era of computer programming, so he described that sort of thing in the mentalistic terms of his day, referring to "categories of thought" that our "minds" impose on the data of our senses.

Interestingly though, when Kant came to the ancient issue of 'substance', he did *not* seem to extend his notion of our obligatory way of thinking to include the things we see around us. So if he were speaking about my canoe, for instance, he might've said: "We know only the *phenomena* (yellow color, scraped appearance, et cetera) manifested by the canoe, but we can never know *the thing in itself.*"

Since Kant was already onto the fact, however, that we can perceive the world only in accordance with our human apparatus, it's a bit surprising perhaps that he didn't take one step further, which would've resulted in his saying: "I cannot even conceptualize without making use of conceptual objects, because that's the only way my information-processing equipment works. For all I know then, there may not even be any underlying thing-in-itself. There may instead be but one overall field of force, varying in intensity. But even if that were the case, my conjectured field-of-force would become simply one more thing in the hands of my information-processing system, given the bedrock way my equipment does its work.

SUBSTANCE AS 'FREESTANDING'

It was within this sort of human limitation that Spinoza examined the notion of 'substance'. His conclusion was that there could in actuality be only *one* substance. Why? Recall that a substance is thought of as free-standing; so my canoe for instance is just sitting there by itself—in marked contrast to its properties (e.g. its yellow color), which have to *inhere* within this freestanding substance. (The term 'sub-stance' derives, not surprisingly then, from two Latin words meaning to *'stand under'*).

Spinoza would've noted, however, that my canoe does not *really* stand on its own. Less than 25 years ago, it didn't even exist; and in the not too distant future, it'll be gone like the wind. There's only *one* thing, he insisted, that literally stands by itself—that's *actually* able to stand under and support all the stuff of our passing scene—namely all-that-is. Importantly, he was not referring here to an inventory analyst's nightmare consisting of each and every item in the universe that happens to be countable at a given time, but rather to what underlies all these shifting 'modes'.

Only this one substance could truly have the reason for its own existence totally within itself; that is to say, depending on no other thing for its existence. How come? Because there was, by definition, no other thing around to give rise to its existence. Consequently, it had to have *within itself* any and all attributes that we're aware of. He then listed the two with which we're humanly familiar—the very two that'd been highlighted by Descartes, namely 'extension' and 'thought' (matter and consciousness).

All the various 'modes' by which this one substance manifested itself could be viewed then under either of these attributes. In other words, extension and thought were to be viewed as two different aspects of one and the same thing. Thus, Spinoza was able to deal with Descartes' insoluble problem of how the human mind was to connect with a human body—with which it shared no common property—by denying that mind and body were two different substances. Rather, they were different attributes of one and the same thing.

Consistent with this notion, he then pointed out that it applied *"not more to men than to other individual things, all of which, though in different degrees, are animated"*[3]—an obvious avowal of panpsychism (and the source of considerable discomfort for modern enthusiasts of Spinoza, given the low repute of this thesis). Sometimes I've wished that Spinoza had focused his brilliant mind more extensively on his notion of panpsychism; but I try not to complain too much, since I've never been enthusiastic about reviewers who grouse because an author didn't write the book they wanted him to write. Spinoza, I remind myself, was writing a book that focused on our human nature; and given that nature, how we could best live our lives (hence the name of his magnum opus, *The Ethics).*

In addition, I remind myself that Spinoza had nothing even close to our modern knowledge of brain function—along with the computer modeling thereof. So when he did mention that the complexity of a thing's equipment would determine the extent of its mental capabilities, his statement was so abstract as to be of little help. In any case, here are the words he used to express the point: *"In proportion as any given body is more fitted than others for doing many actions or receiving many impressions at once, so also is the mind of which it is the object, more fitted than others for forming many simultaneous perceptions…"*[4]

The good news is that with our modern knowledge we'll be able in the coming chapters to put some flesh onto that vague statement. Note for now, however, that the procedure of taking irreducible elements and kicking them back to the git-go is exactly the procedure used by modern particle physicists. That's in effect what Spinoza did also, and that's what we'll be doing too. First, however, it'll be worth our while to see how Spinoza's analysis <u>naturalized</u> what had previously been viewed as supernatural, a step that's significant for us to clarify, since our present account is to remain totally within a Naturalist world view.

28

NATURALISM

Readers who've taken the time at some point in their busy lives to study proofs for the existence of God probably had feelings of *deja vu* while reading the account of Spinoza's insistence that there could be only one substance. That's because he pretty much took the old Scholastic proofs of God and applied them innovatively to his notion of substance. That is to say, the God ('Being from itself') of the Schoolmen, just like Spinoza's substance, had to have *from within* all that was necessary to support its own existence. Perhaps not surprisingly then, Spinoza referred to his one substance as "God." The way he expressed it in the Latin of his time was *"Deus sive Natura,"* which translates to "God or Nature."

Why not a Transcendent Being plus all the substances we see around us that "He" has created? Well, recall that Spinoza's take on the situation was that there can actually be only one substance—all other things then amounting to 'modes' of this one substance's existence. So in that sense, there's nothing outside of Nature. Restated for emphasis, whatever *is* exists within Nature *(Deus sive Natura)*.

Let's contrast this view with the more traditional view of God as outside and above the natural stuff of the universe that "He" has created. Descartes of course had maintained this traditional view, so his world consisted of various substances that existed separately from the Substance of God. These subsidiary substances included the material substance of our bodies and the non material substance of our souls. Congruently then, things like our height, weight, and skin color were to inhere in our bodily substance; while things like consciousness and conceptualizing were to inhere in our non material mind substance.

If, however, there are no other substances than the one actual substance—no other things that can really stand by themselves in order to support the properties said to inhere in them—then there cannot be other created substances, but only 'modes' (as in 'modifications') of the one true substance. And once again at this point, Spinoza was to make innovative use of a longstanding thesis about God.

The setting in which this thesis had originally developed is interesting enough that the story may be worth relating.

GOOD GOD, BAD GOD

An ongoing problem for believers in the God of Abraham, Isaac, and Jacob, had to do with the need to explain why all sorts of evils abound in a world created by this All Good Deity. And considering the very human traits that were being attributed to this Deity, one plausible answer popped readily to mind—a scenario of the guy with the white hat versus the guy with the black hat. This sort of depiction gave rise in the third century C.E. to a temporarily prospering Religion called Manicheanism (even attracted the likes of Saint Augustine for a while). The Manichean 'heresy' featured a Good God *and* an Evil God, the two being locked in eternal combat. We humans, possessed of both noble spirit and carnal body, were caught in the middle, so to speak, and our job was to choose the spiritual part of ourselves while conquering the evil of our bodies.

That in turn involved eschewing sex (a stance that never bodes well for an Institution's long term health), and while Augustine remained pretty enthusiastic about this particular eschewment even after his conversion to Catholicism, more practical heads within the Church eventually prevailed. To some degree. But at any rate, to get on with our story, Scholastics pointed out that the notion of two gods is logically inconsistent. For if there is an *Ens a se*—a Being who has all that is needed for its existence within itself—then this Being cannot be limited by something outside itself like another god. But since Spinoza took the term 'substance' in a similarly strong sense to mean that which can actually exist by itself, then this substance, likewise, could not be limited by other substances that were proposed to exist outside itself.

DESCARTES' SUPERNATURALISM, SPINOZA'S NATURALISM

Let's look now at how these two views play out in the case of our own grand and glorious species. Descartes, as we've seen, developed a two substance theory of man—material body and non material mind—and he used these two substances as separate containers for different properties. But within this theory, how are we to explain the manner in which our species propagates itself?

Clear enough that children come by their bodies via egg and sperm cells they've inherited from mom and dad. But what about their *mind substance?* What sort of non-material-DNA equivalent might be passed along to them? After all, one of the core characteristics of Descartes' proposed soul substance was that it can *not* be divided into parts (Chapter 22); so it's not as if some piece of a parent's soul could separate, thence to join up with sperm or egg in order to pass immaterial souls along to the next generation.

Fortunately, Descartes didn't have to worry about this problem too much, because everyone within his culture already *knew* how this was to come about; namely, that God would "infuse" a new human soul into the fertilized egg at the moment of conception or not too long afterwards. This in fact is why the act of human generation got referred to as "procreation." Parents are to supply sperm and egg, but the Transcendent God has to intercede from outside this natural process and create a spanking new human soul.

But since Descartes' two-substance thesis requires special creation of this sort to account for each and every new human being, his explanatory story ends up being Supernaturalist at its very core. Alternately stated, his theory requires miraculous interventions on a daily basis. In marked contrast, Spinoza's one substance *is* Nature; hence all the modes by which it manifests itself are within nature, including therefore both the attributes of thought and extension (consciousness and matter).

While our modern ways of expressing things—along with our greatly enlarged information base—relegate Spinoza's manner of expressing himself to a bygone era, I've referred to it at some length here because it's so congruent with our modern scientific worldview in so far as it focuses on explanations that flow from *within* Nature.

And if, by the way, some readers are surprised at our interjecting the notion of God into the present Naturalist account, we should note that to do so follows pretty readily within any discussion focusing on what exists within nature from the git-go. That's because any notion of what exists from the beginning overlaps conceptually with the conventional notion of *God*—albeit Spinoza's use of that term was so alien to the folks of his time that he was often labeled an atheist (indeed he was excommunicated from his Jewish community in Amsterdam).

ALL THAT'S WITHIN NATURE IS NATURAL

I've been using Spinoza's account of Nature here in order to provide historical background for a notion of *consciousness-as-fundamental*. Nonetheless, we should keep in mind that my primary patrons in this regard consist of our much revered particle physicists. Their *modus operandi* consists in breaking down complex items into progressively simpler components. But when they reach rock bottom, so to speak, they then treat what remains as fundamental to nature. So, for instance, the particle physicist's current *"Standard Model"* consists of 12 particles and 4 forces. These then are treated as fundamentals. However, these factors don't necessarily encompass *all* of the fundamentals. That's because "physics" deals with "the physical," so physicists leave consciousness outside of their formal considerations. Best that they can hope for, as Nobelist Steven Weinberg expressed it in his *Dreams Of A Final Theory,* is *"that we will come to understand the <u>objective correlatives</u> to consciousness in terms of physics (including chemistry) and that we will also understand how they evolved to be what they are...That may not be an explanation of consciousness, but it will be pretty close."*

As we've seen in Chapter 25, however, correlation of human consciousness with human brain function doesn't explain the occurrence of consciousness, and to rest on a nebulous statement that consciousness *"supervenes"* on certain complex information-processing activities provides little actual advance in our understanding. We seem at present then to have two options: Either we take consciousness to be one of the fundamental natural ingredients of our universe, or we say that consciousness supervenes under certain conditions of complexity, even though we're at a total loss to explain how this occurs (c.f. McGinn's view, summarized in Chapter 25, that our information-processing apparatus is "cognitively closed" to comprehending the necessary link).

NATURE ACCORDING TO THE 'MATERIALISTS'

Before opting for the acceptance of consciousness as a basic property within nature, we should, however, in the interest of...*uh*...fair and balanced reporting, look further into the perspective of *philosophical materialists.* While doing so, it's important first of all to distinguish these individuals from the "materialists" of our everyday speech—those spiffy folk who place an overarching value on the accumulation of fancy belongings. In wretched contrast, philosophical materialists, alas, often consist of no more than pathetically impecunious academics who come by the name 'materialist' only because they erect their vision of reality on a platform provided exclusively by matter and its movements.

Since ancient times, such a thesis has appealed especially to hard headed, no nonsense, types wanting to set their theories on *terra firma.* Still though, one of human history's most poetic philosophers, Titus Lucretius Carus, espoused the approach quite fervently—and this time I'm referring literally to a "poet of philosophy," because Lucretius rendered a lengthy account of his thoughts in real honest-to-goodness poetry during the first century B.C.E. What he came up with was a work of sufficient beauty that readers can still find translations available on bookstore shelves *(On The Nature Of The Universe)*.

At any rate, after his de regueur appeal to the gods for guidance, Lucretius proposed to *"reveal those atoms from which nature creates all things…Or I may call them 'primary particles', because they come first and everything else is composed of them"*[2] (translation by R.E. Latham). Those of course were the good ole days for materialists, before it was discovered that a-toms (not-cuttables) were filled mostly with empty space, and that atoms were indeed eminently cuttable. Furthermore, even if some of their constituents like electrons appear to be not additionally divisible, these building blocks turn out to be as much like waves as particles. In short, the once *firm ground* of matter seems to have vanished underfoot.

But current materialists have not been dismayed. And in fact they seem to revere modern particle physicists even more than myself, because materialists now start with the revised assumption that *"[the 4-forces-and-12-particles] come first and everything else is composed of them."* Alternately stated, the 4-forces-and-12-particles underwrite all that is natural; hence, any basic attribute outside of these building blocks is not to be spoken of as "natural." Obviously, if one starts with such an assumption, then it follows like the night the day—only more so—that the explanatory model I'm beginning to run by you does not provide a "Naturalist" account.

OCCAM'S RAZOR ONE MORE TIME

Before making the obvious rejoinder that starting with this assumption narrows the playing field arbitrarily along the line of one's prior preferences, we should note that—in accordance with the natural proclivities of our limited information-processing systems—we have an almost overwhelming tendency to favor simplifying explanations. So if our Standard *4-force-12-particle* Model, operating at highly complex levels of interaction, were up to the task of explaining the supervenience of consciousness in understandable fashion, Naturalists like myself would be grasping the explanation with simple (not to mention simplifying) joy. That's of course in keeping with the Law Of Parsimony as described in Chapter 26.

Only when such an effort has failed then—or at least has not succeeded—do we feel pushed to look for alternatives. However, since most theorists are loathe to relinquish their most basic tenets, if one is wedded to a *materialist model* of what's natural, then it's hard not to insist that one's conceptual model has indeed been up to its explanatory task. We've already seen in Chapter 26 a couple of ways in which theorists continue this insistence, but there's an additional argument often used, and it's one that I find intriguing.

THE BURDEN-OF-PROOF ARGUMENT

Here's the background: In any endeavor where immediate action must be taken, one is *forced* to make a decision. This is most famously the situation in courts of law, where a practical result needs to be forthcoming, as in "You are found guilty and must pay the penalty," or "You are found not guilty, so you're free to walk." In any such case of forced choice, it helps when one of the parties in conflict is required to assume the burden of proof.

In United States law, where one is considered "innocent till proven guilty," the burden of proof falls on the Government. So, for instance, when John Hinkley shot President Reagan, defense lawyers claimed that their client was insane, while prosecutors insisted he was indeed responsible for his actions. Since the resolution of this disagreement had important practical implications, it was useful to make one side assume the *burden of proof.* Hence, once defense lawyers produced evidence to support a plea of insanity, it was up to the prosecution in effect to prove that Hinkley was sane.

Now let's examine how materialists want to use the burden-of-proof argument to further their cause. "Look," they say, "we have solid scientific evidence that the basic 4 forces and 12 particles exist. Furthermore, all the familiar objects of our daily experience result from complex configurations of these basic building blocks at various levels of complexity. We now understand, for instance, that what we refer to as 'life' involves highly complex alignments of basic constituents in a fashion that permits self-replication. In like manner, complex information-processing systems formed by the most complex of these replicators result in what we refer to as consciousness."

Objecting theorists then reply: "That's what bothers us. We don't see how any ilk of information-processing necessarily leads to or amounts to consciousness, so something's missing here."

In turn, the more polemical type of materialist responds: "You're trying to make <u>me</u> assume the burden of proof here, requiring <u>me</u> to convince you that a complex information-processing apparatus like our nervous system will be marked by consciousness. But I think it's time that <u>you</u> assumed the burden of proof. How about <u>your</u> proving to me that this sort of apparatus can function *without* producing the consciousness that we know comes with it?"

My legalistic side really warms to this argument, which is why I mention it. But of course the fact is that theorists are well known to live in Ivory Towers, and usually their theoretical judgments do *not* require forced choices. Quite often then, they simply remain agnostic. In this case, for instance, a sensible rejoinder might be: "Well, it's possible of course that your conclusion might be correct, but unfortunately your arguments are less than convincing. That's why I'm taking the time to explore other explanations that seem to me to make good sense."

At this point, materialist philosophers usually throw in their ace of spades: "Good sense?!! Why, the explanation you want to pursue, namely panpsychism, is so *patently absurd* that it shouldn't even be dignified as a serious explanatory effort." Time now to examine the issue of whether panpsychism is guilty as charged.

29

PANPSYCHISM AND ANIMISM

When I was a little boy, one of my favorite rhymes was about a Gingham Dog and a Calico Cat that side by side on the table sat. Next morning, as the story had it, the dog and cat were both missing. So what'd happened? Seems the two of them got into an awful argument and ended up in a fight of such ferocity that they ate each other up.

Let's forget about the faulty physics here for a moment and concentrate on the *animism* that's involved. Little children readily project human scenarios onto inanimate objects, so it was no problem at all for me to imagine these two stuffed dolls getting into a heated argument and having a fight (just like me and my brother, John). That's of course because when I was a child I thought as a child. But now that I'm an adult, although the rhymes and the clever ending still provide amusement, the notion of arguing dolls, taken literally, would strike me as *patently absurd* (if I may be allowed the privilege of using a favored philosophical putdown).

Under the circumstances, I was a bit astonished to read in Oliver Wendel Holmes' famous old treatise on *The Common Law* that many grown up folks during Roman times seemed to take such animist views quite seriously. Here's Holmes' account of one such circumstance, occurring of all places in courts of law:

> As late as the second century after Christ the traveler Pausanias observed with some surprise that they still sat in judgment on inanimate things in the Prytaneum. Plutarch attributes the institution to Solon…In the Roman law we find the similar principles of the *noxae diditio*…The action followed the guilty thing into whosoever hands it came…[1]

Holmes commented:

> But it may be asked how inanimate objects came to be pursued in this way, if the object of the procedure was to gratify the passion of revenge. Learned men have been ready to find a reason in the <u>personification of inanimate nature common to savages and children</u>…Without such a personification, anger toward lifeless things would have been transitory, at most.[2]

Yet the cultures of ancient Greece and Rome were hardly operating at the level of mere "savages and children," so the practical power asserted by anthropomorphic projection in these relatively advanced cultures provides a good illustration of this mental mechanism's pervasive strength. Of course, we've made considerable conceptual progress since that time. Nowadays, outside of poetic metaphor and…*uh*…the naming of hurricanes, almost no educated person would indulge in the animistic personification of inanimate objects.

Now a pertinent question arises: Does the notion of *panpsychism* necessarily involve *animism* (the personification of inanimate nature)? Because, if it does, we'd have no choice but to scrap panpsychism as a serious theory. There's no question, historically, that these two concepts *were* usually run together. Not only that, when current arguments on the subject arise, these two concepts still tend to get conflated. Let's illustrate that fact now by taking ring-side seats at a fight between two top heavyweights among current theorists of mind, the old pro, John Searle, versus the young Turk, David Chalmers.

SEARLE VS. CHALMERS

Chalmers had shared with readers of his influential book, *The Conscious Mind,* an account of his initial sortie into the theory of mind. Turns out that he'd attempted at the beginning to maintain the received wisdom about consciousness supervening on highly complex information-processing systems, and only after he became dissatisfied with this attempt to draw the rabbit of consciousness out of the hat of ordinary supervenience did he start looking for something additional. In the process, he ended up giving serious consideration to the notion of consciousness as a fundamental property of nature—or in his words, *"to admit phenomenal or protophenomenal properties as fundamental."*[3]

When John Searle took on the task of reviewing Chalmers' book, Searle minced no words. As he put it: *"Of all the absurd results in Chalmers's book, "panpsychism is the most absurd and provides us with a clue that something is radically wrong with the thesis that implies it."*[4]

Searle presented absolutely no support for his brusque assertion of absurdity—obviously because he judged that none was called for. He simply assumed that any theory leading to such a result produced its own *reductio ad absurdum*. But Chalmers just shook off the blow, and in a later exchange, he noted: *"In place of substantive arguments, Searle provides gut reactions: every time he disagrees with a view I discuss, he calls it 'absurd'."*[5]

Seeing how well his opponent could fight back with a counterpunch to his own gut, Searle then tried some shots to the head. Or at least he decided to address Chalmers' conceptual notions head on:

> We know that human and some animal brains are conscious. We also know that consciousness in these systems is caused by quite specific neurobiological processes…Now, for someone seriously interested in how the world actually works, thermostats, rocks, and electrons [Chalmers' examples] are not even candidates to have anything remotely like these processes, or to have any processes capable of having equivalent causal powers to the specific features of neurobiology. Of course as a science-fiction fantasy we can imagine conscious thermostats, but science fiction is not science. And it is not philosophy either.[6]

We'll pass over the fact here that we don't really *know* what Searle thinks we know, because it simply isn't certain that "specific neurobiological processes" cause consciousness. If, for instance, a Spinozist type of <u>dual aspect theory</u> turned out to be the best explanatory story, then what we perceive on the one hand as our brains at work, and on the other hand as our accompanying states of consciousness, would be simply different aspects of the same thing when viewed from different perspectives. One would not be causing the other. Additionally, many theorists reject Searle's opinion that "specific neurobiological processes" are required, suggesting instead that a computer-chip system, properly constructed to capture the information-processing functions of a complex brain, would equally be associated with the occurrence of consciousness.

Reason to pass over this fact quickly is that we're focusing at the moment on our central point, namely, that theorists so often run the notions of panpsychism and animism in together as if they were inherently the same, which they are not. Recall, for instance, Spinoza's abstract statement noting that *"In proportion as any given body is more fitted than others for doing many actions or receiving many impressions at once, so also is the mind of which it is the object, more fitted than others for forming many simultaneous perceptions…"* Let's specify Spinoza's statement a bit now by focusing on the fact that some bodies are much more fitted than others for performing highly complex computational activities—including layer after layer of additional computations performed on results initially obtained. This is the sort of activity that allows an organism possessed of an extremely complex information-processing capability to *represent* itself as the one seeing, moving, or thinking (more even, because it allows the organism to represent itself as thinking about the thinker who was just thinking about how this thinker thinks).

CONCERNING CONSCIOUS THERMOSTATS

But to drive home the point that panpsychism and animism are not inherently the same, let's try to perform a thought experiment that Searle apparently thinks we can do pretty readily *("Of course, as a science-fiction fantasy we can imagine conscious thermostats...")*

Before starting on our venture though—with me as the guinea pig—let's recall that science-fiction aficionados make quite a distinction between this category and that of fairy tales. Harry Potter stories, for example, are magical (i.e. outside of nature), not science-fictional. Science fiction attempts to take what we scientifically know and extrapolate its plausible consequences into some future time or into some novel situation. Purist fans of this genre get turned off mighty fast if a proposed science-fiction story indulges in the magical breaking of known rules of science.

What I have to do here then is to try imagining a conscious thermostat while remaining within the dictates of science. So if we are to apply a dual aspect hypothesis, we need to have the relevant systems up and running in order for appropriate conscious manifestations to occur. To illustrate what happens if relevant systems are *not* available, we need only recall that if human information-processing systems underwriting language are destroyed, our consciousness will not extend to awareness of language (as happens, for instance, with stroke patients diagnosed as aphasic).

For a warmup now, let me contrast my human apparatus with that of the rocks mentioned by Chalmers and Searle. First of all, I have highly complex detector systems, while rocks have none at all (in any *organized* way at least, though of course a rock will, in its extremely primitive manner, register, say, the heat of the sun). And rocks also have no information-processing systems to make sense of the ambient energy detected, or motor systems capable of responsively adaptive behavior. Hence, I'm left with almost nothing for my human imagination to legitimately work with.

Let's move on then to thermostats. Since this instrument *does* have a simple detector system for registering ambient temperature, and since it processes the information gleaned in a way that leads to a decision (on/off), and further, since it's then capable of flicking an action switch, there are at least a few things happening that I can start working on in my imagination. Nevertheless, I still have to be very careful if I'm not to abandon science-fiction and resort to magic. I'll try not to imagine, for instance, what the room I'm in looks like, because thermostats don't have vision.

The only thing thermostats sense is warm or cold, so I imagine myself now getting cold. And *that* in fact permits me a humanly happy comparison, because 'cold' for a thermostat depends on its setting. Likewise for myself, 'cold' so often depends on my own system's setting. For instance, if my hand's set in hot water and then suddenly placed in tepid water, my detectors will sense 'cold'. In this context, I as thermostat now sense 'cold' and send a message to my effector mechanism (I as thermostat have only one) in order to close the circuit that starts the furnace.

And now I have still another happy comparison. For if the real I tells myself to flick the furnace switch on, I don't have any conscious experience of how this decision has gotten my arm to move toward the switch and my right index finger to flick at just the right moment. In like manner then, I-as-thermostat have no need to consciously experience the fact that two attached metals of unequal expansiveness in my apparatus are bending in a way that closes the circuit.

I (the real I again) then finish my task of trying to imagine a conscious thermostat by subtracting as many additional features of human consciousness as I can. For instance, the thermostat has no mechanisms available within itself for evaluating the quality of its performance or for subsequently experiencing pleasure, or disappointment, or boredom. So it's not up to thinking anything like "That's a good job, and you should feel proud of yourself." And in any case, it can have no sense of self because the thermostat has no systems with which to monitor its lower level systems. And of course it has no linguistic apparatus, so I've got to try my best to avoid adding any verbal elements like 'cold', 'self', and 'performance'—even though that's nigh on impossible for me to accomplish because words seem to follow my conscious activity as inexorably as a vapor trail flows in a 707's wake across the morning sky. Nearest I can get is to keep reminding myself to block out all verbal elements, but the reminders themselves just introduce more words.

To summarize, the best I can do when attempting to legitimately imagine a conscious thermostat is first to project my human consciousness and then try to subtract as many of its sophistications as I can. Yet try as I might, I always end up administering a heavy dose of anthropomorphic projection. Bottom line: I'm not really able to come up with a science-fiction fantasy about what a conscious thermostat might experience. To do so, I'd have to switch to magical thinking. Then of course I'd be able to tell the story of a rebellious thermostat, or a helpful thermostat, or whatever other human trait I decided to project onto my fictional thermostat.

But there are after all many scientific constructs that we're unable to reproduce in our imagination. Chapter 11, for instance, noted the frustration I experienced as a child while trying desperately to picture Einstein's four dimensions. Problem with that effort was that *our computational equipment obligatorily breaks the space-time continuum into mid size objects that change over time, so members of Homo Sapiens cannot directly depict all four dimensions together.* Einstein was no more able to visualize the situation directly than I. What he was able to do was to express the concept (in that particular case with quantitative precision) by using abstract mathematical notations.

The analogy here is that we can't truly imagine <u>any</u> consciousness except our own human consciousness (with perhaps one exception that we'll get to later). When another animal's cognitive equipment is close enough to our own, we think we can get *some* notion of what its consciousness may be like. But when we can't get even close, we generally decide that this type of organism is *not* really conscious at all.

Now, if being "conscious" means to have the sort of awareness that we can get close to imagining—at least a little bit—then we're absolutely right to deny "consciousness" to the most primitive creatures. In that case, however, in order to maintain the plausible hypothesis of consciousness as fundamental, we need to come up with another word, and 'protoconsciousness' has sometimes been used for this purpose.

But in terms of *imagining* what such protoconsciousness might be like, don't even waste your time trying. We might as well attempt to imagine the 11 dimensions proposed by Superstring physicists! In both cases, we need to be satisfied with abstract concepts that are not directly translatable into products of our human imagination. Or to pursue the point with yet another analogy, we might say that just as gravity requires a certain mass in order to become recognizable as a force, protoconsciousness would require a certain mass along with certain conformations in order for us to attribute "consciousness" of a humanly recognizable sort.

CONSCIOUSNESS TO US IS HUMAN CONSCIOUSNESS

Crucial to my explanation above is the notion that consciousness, in any way *directly experienceable* by us, is human consciousness. That being the case, when performing exercises of our imagination, we have no choice but to project our human consciousness onto whatever's at issue. So we always end up, practically speaking, with some degree of anthropomorphic projection. But if consciousness is to be conceived as something far broader than our own version thereof—something that's capable of extending as far as the fundamental ground of all that is—we have to give up trying to envision it directly, except as an abstract concept. In order to hammer that point home, it'll be useful now to illustrate that fact in both upward and downward directions. With that task in mind, we'll start by examining one of the most riveting instances of our anthropomorphic projection.

30

GOD'S CONSCIOUSNESS

One of my favorite bible stories (at least from the perspective of gallows humor) involved some earnest negotiations between Abraham and the Lord. Many readers will be generally familiar with the story of Sodom and Gomorrah, where unrepentant inhabitants evoked their own destruction as a result of God wreaking Divine Justice upon them. But some readers may not recall that this happened only after Abraham had done his best to prevent…*uh*…excessive collateral damage. Or as he was to put it to the Lord:

> Wilt thou destroy the just with the wicked? If there be fifty just men in the city, shall they perish withal? And wilt thou not spare that place for the sake of the fifty just, if they be therein? Far be it from thee to do this thing, and slay the just with the wicked…This is not beseeming thee…And the Lord said to him; if I find in Sodom fifty just within the city, I will spare the whole place for their sake.[1]

And though Abraham had obviously done quite well in this opening round of negotiations, he was still a bit uneasy, probably because he was wondering whether he'd actually be able to round up that many good men in Sin City. So he proceeded to ease his way into a slippery slope argument:

> What if there be five less than fifty just persons? Wilt thou for five and forty destroy the whole city? And he said: I will not destroy it, if I find five and forty. And again he said to him. But if forty be found there, what willt thou do? He said: I will not destroy it for the sake of forty. Lord, sayeth he, be not angry, I beseech thee, if I speak: What if thirty shall be found there?[2]

And continuing to maneuver in this manner, Abraham—whom I would nominate as the patron saint of all negotiators—succeeded in edging the Lord down to twenty and finally down to just ten good men. Now, there are many different ways in which this story could be analyzed,[3] but our present point is that God is viewed here in highly human terms. Powerful, that's for sure, but like some good boss that we may've come across in our own personal experience: What he says goes, but at least he'll listen to what we have to say; and if it makes sense to him, he'll be influenced by it.

Problem is that the *God-of-the-philosophers* can't really fit into this sort of good-boss scenario. For when the Scholastics reasoned to the existence of a Being they called "God," that Being (*Ens a se*) was to have, through logical necessity, *"every perfection to every degree of perfection."* We humans, highly limited as we are, start our ventures off with an opening game plan—if we're sensible—and then we make appropriate alterations as we go along. But a Being who is All Knowing and All Powerful doesn't plan out future contingencies, and He doesn't change his plans. In fact, He doesn't *plan* at all, because everything's present to Him in the first place. Only limited beings like ourselves plan—and need to alter our plans when we find reason to change.

In starkest contrast, the God of the philosophers is <u>totally not human</u>. Yet when people *imagine* God, they cannot help but envision Him as very human indeed—only that He's much more gloriously so. They imagine *Him*, that is, as the Superest of Supermen. And I italicize *"Him"* because Christians, following a patriarchal tradition antedating Christ by centuries, thought of God in *'man'* terms, specifically as a father rather than a mother.

Since Spinoza tried to be logically consistent at all times, he did his best not to project human traits onto "God." Here's how he put it:

> ...since theology frequently, and not unwisely, represents God as a perfect man, it is often expedient in theology to say, that God desires a given thing, that He is angry at the actions of the wicked, and delights in those of the good...[But] in the language of philosophy, it cannot be said that God desires anything of any man, or that anything is displeasing or pleasing to Him: all these are human qualities and have no place in God.[4]

Let's illustrate why this is so by contrasting the Divine situation with our own: <u>Desiring</u> stuff is pretty common for us, because we so often lack things that'd be nice to have. But if a Being already has *within itself* everything it needs for the fullness of its own existence, then there's nothing left to be desired. Stated from the other direction, desire implies something that's *out there* to be desired—something that we don't already have, or have enough of.

Likewise, we are often <u>pleased</u>. That's because even our happiest moments are relative in degree, so something can always come up that'll make us even happier. But once again, if a Being already has *within itself* absolutely everything it needs—"every perfection to every degree of perfection"—then there's no room left, so to speak, for any additional pleasure. And of course, in the other direction, there's no possibility of being displeased, a state that implies having been deprived of something.

And we often plan what we're going to do, designing along the way things that'll be helpful in attaining our goals. But as Spinoza noted: *"in eternity there is no such thing as when, before, or after...God [in Spinoza's sense of that term] did not exist before his decrees, and would not exist without them."* That being the case, in pure logic, God does not 'plan' or 'design' anything.

Ironically, it's in large measure because Spinoza did not envision the "Infinite Consciousness" of God in a highly human manner that so many people looked on him as an atheist. However, it's not just Jews and Christians who think of God as some sort of "perfect man." The key point here is that when any of us tries to think of another being as conscious, we have no choice but to envisage that consciousness in terms of our own human consciousness. Not only do religious folks treat the proposed consciousness of God in this way; so do hard-headed scientists who champion a Naturalism made possible by current evolutionary theory. We can illustrate this fact by looking at a current controversy between religious Fundamentalists and Evolutionists.

EVOLUTIONISTS AND THE HUMAN *NOTION* OF 'GOD'

First we need to keep in mind that before the thesis of Natural Selection became available, "Creation by intelligent design" was the only show in town. Even the forceful Voltaire—he who spent a lifetime railing against the Institutional Religion of his time—had to acknowledge the hand of a Superior Being in order to account for the purposefulness he observed within nature. *"God,"* he said, *"gave bees a powerful instinct which makes them work and feed together, and he gave man certain feelings which he can never shake off...Our benevolence toward our own species, for example, is born with us and is always working within us..."*[5]

The 19th Century British preacher, William Paley, had eventually provided the most popularly moving account of this 'design' explanation, which went something like this: If you were walking in the woods and came across a watch lying on the ground, you would immediately conclude that this intricate instrument had been made by someone of high intelligence. Now look around you at the plants and animals. In similar fashion, one must conclude that these far more intricately designed instruments have also been purposefully fashioned, and by an intelligence infinitely superior to that of a mere watchmaker.[6]

This then was a modern version of one of the Scholastic proofs for the existence of God that had been called the "Argument from Design." But once Natural Selection became explanatorily available, the "God of the gaps" could once again be dismissed. To that end, a persuasive new preacher of Naturalism, Richard Dawkins, wrote a best seller entitled *The Blind Watchmaker*[7]—his title being both a clever play on words and a neat analogy, because Natural Selection required no insightful planning. Instead, a combination of genetic mutations and adaptive selection would get the job done.

WHAT ABOUT SOPHISTICATED BELIEVERS IN A TRANSCENDENT GOD?

Easy to see then why religious Fundamentalists, devoted to a humanoid god, got up in arms—and are still fighting a fierce rear-guard action against Natural Selection to this very day. But what about sophisticated believers in a Personal God? Here's the way they look at things: "Don't you think it's more than a bit coincidental that the laws of nature turn out to be such that replicators (i.e. living organisms) can form in the first place? And then that the utterly ingenious mechanism of Natural Selection is able to take it from there?" They continue: "One has only to imagine slight changes in our natural laws—a minuscule difference in the size of Planck's Constant for instance—and the whole world as we know it could not have developed. So scrap if you will the analogy of God as mere artisan. Let's think instead of an Intelligent Designer who's infinitely more sophisticated, readily capable of formulating a plan that causes nature to unfold spontaneously in all its beauty after having once been set on course."

OUR IMPROBABLE WORLD

That's an argument not to be sneezed at—God as Super Robotics engineer—especially for those who think it's perfectly congruent with logic to posit a "First Cause" that's *outside* of Nature. Scientists of a Naturalist bent do, however, have a response to this probabilistic argument; namely, that if the Laws of Nature had turned out to be different—if other laws had come on line instead—then we wouldn't be here to be puzzling about things. So we have to start with the fact that this *is* the World that's actually here, no matter how statistically improbable.

And as a matter of fact there are lots of incredible longshots that happen, though ordinarily we take them for granted. What was the likelihood in each of our own cases, for instance, that our forbears would come across each other in the first place, even just a few generations back? And then what was the likelihood that out of the many millions of sperm our great great great grandfathers produced, that the exact ones would be selected to produce our particular great great grandfathers? Continue to *multiply* those huge odds then across each subsequent generation (not even bothering with all the other unlikelihoods) and the existence of any one of us is virtually miraculous. In like manner, while the occurrence of our own special Universe with its own special laws may be statistically improbable, we *are* in fact here; and from *that* starting point, our improbable selves can't help but ask: "What's it all about, Alfie?" But of course if things had been otherwise, there'd be no one here to ask such questions.

Now, while the above debate is interesting in itself, I summarize it primarily to illustrate that both sides agree: **_either_ there was a Designer, _or_ that what's happened is a matter of chance**. No third alternative. At first glance, by the way, this dichotomy seems to logically favor those who reject the notion of design. That's because matter-of-chance theorists can readily note (*a lá* Spinoza) that *"God did not exist before his decrees,"* so it's not as if there were some humanoid God out there saying to himself: "Now let's see, what sort of stuff would be good for me to create in order to demonstrate my glory?" So why not simply speak instead of regularities that we observe in nature, rather than invoking a personalized consciousness who's planning things out?

What's appealing about this view is that a focus on *regularities* avoids personifying God, in marked contrast to the notion of a Personal God, which cannot fail to project images of our own human consciousness. And the fact is that we can scarcely begin to imagine, or conceive, how the fundamental attribute of consciousness would manifest itself in the one substance that underlies all individual modes. Nevertheless, to allow only one alternative to the notion of design—namely the alternative of brute regularities interacting by chance—may demonstrate no more than another instance of our limited cognitive apparatus at its work. For restricted as we are to the experience of our own human consciousness, we may well be incapable of coming up with the relevant category to fit this situation. That's because, while it's correct to say that nothing existing in *Deus sive Natura* has been "planned," nonetheless the attributes of that one necessary substance hardly seem to squeeze into the category of "mere chance" very well either.

HELP OF THE HELPLESS, OH ABIDE IN ME

Before leaving the subject of what happens when we try to project our human consciousness 'upward', perhaps it would not be amiss to note that it's virtually built-in for us to *want* a Superman God. For whenever there seems to be no solution to a horror that we're facing, it's hard *not* to wish for a personally attentive God to reach down—like a father with his small child—to soothe our pain. Seems to me, in fact, that there were rippling eddies of this sort of sentiment even in the no-nonsense analysis Nobel physicist Steven Weinberg provided for Einstein's judgment concerning the God of Spinoza:

> Einstein once said that he believed in "Spinoza's God who reveals Himself in the orderly harmony of what exists, not in a God who <u>concerns himself</u> with fates and actions of human beings." But what possible difference does it make if we use the word "God" in place of "order" or "harmony," except perhaps to avoid the accusation of having no God?[8]

Then Weinberg went on to ask: *"Will we find an <u>interested</u> God in the final laws of nature?"* Why bother to ask that particular question, I would wonder, unless one has at some point entertained this humanly comforting notion? At any rate, Weinberg then answered: *"All our experience throughout the history of science has tended in the opposite direction, toward a <u>chilling</u> impersonality in the laws of nature."* I've underlined the humanly chilling word *"chilling"* because that's what it sure seems like whenever we're in a help-of-the-helpless state. Notwithstanding our psychological neediness, however, we don't inherently need to convert "God or Nature" *(Deus sive Natura)* into a personal helper of humanity in order for us to experience awe, reverence, and deep pleasure. Witness Spinoza himself, who was so utterly enthusiastic about his non-Superman God that one of his more famous commentators went so far as to say Spinoza was "God intoxicated."

PLEASURE AT OUR CONTEMPLATION OF WHAT IS

And on the subject of "deep pleasure" that some folks feel in the contemplation of 'What Is', let's attend a moment longer to Weinberg's dreamings of a final theory. After noting that the strongest motivation behind belief in an *"interested God"* has not been the urge to speculate about *"infinitely prescient first causes"* but rather has to do with our human longing for a string of happy interventions by such a Being, he continued:

> I have to admit that sometimes <u>nature seems more beautiful than strictly necessary</u>. Outside the window of my home office there is a hackberry tree, visited frequently by a convocation of politic birds: blue jays, yellow-throated vireos, and, loveliest of all, an occasional red cardinal. Although I understand pretty well how bright colored feathers evolved out of competition for mates, it is almost irresistible to imagine that all this beauty was somehow laid on for our benefit. But the God of birds and trees would have to be also the God of birth defects and cancer.[9]

I underlined the part about the beauty of nature because it provides us with an opportunity to revisit the issue of esthetics that we remarked on briefly in Chapter 20. When viewed from within a context of *Comparative Cognition,* one of the first things to note is that different organisms find different stuff attractive. The 'skunk cabbage' growing in my marshland is well named—from my perspective—because I find its aroma dismal. Yet a certain type of fly is attracted to its grungy flowers as if it were imbibing nectar of the gods. And on the subject of flies, whenever I see two of them mating, I wonder how such ugly creatures (to me) can be other than repelled by their opposite-sexed counterparts. Yet if it's true that we vote best by our actions, the fly's sense of attractiveness obviously differs markedly from my own.

Nevertheless, all organisms *do* end up with one or another set of esthetic weightings. As for our own species' sense of beauty, although it may not be immediately clear why a given item evokes our esthetic pleasure, I doubt that nature is really *"more beautiful than strictly necessary."* For as noted in Chapter 20, "Our sense of beauty—our esthetic sense—may not be too far behind [our moral sense] in importance, because the pleasure it provides 'helps us bear those ills we have, rather than rushing to others that we know not of'." If members of our species were not hard-wired to find so many things attractive in life, I wonder how many of us would hang in there through those crunch times when we're nearly overwhelmed by life's pain. Seems to me in fact that under even the ordinary "one-durn-thing-after-another" conditions of human life, nature is just beautiful enough to keep most of us hanging around.

At any rate, so much for our attempt to envisage consciousness in an 'upward' direction. Let's look next at what happens when we try to imagine the consciousness of our fellow creatures on Spaceship Earth.

31

WHAT IS IT LIKE TO BE A BAT?

Academicians sometimes gauge a fellow theorist's influence by counting the number of times his work gets cited in the professional literature. And I use the masculine pronoun here, not just because of my male-chauvinist upbringing, but also because the theorist I have presently in mind is Thomas Nagel, whose 1974 paper *"What Is It Like To Be A Bat?"* still gets widely referenced by theorists of mind after all these years. And like many classic articles, its sphere of influence has extended to many who've never had occasion to read the original. That included me a while back, which partially accounts for why I was so quick to reject Nagel's contention that a bat's echolocation isn't something we can humanly imagine.

Here's where I was coming from: One of my pleasant memories from childhood was of accompanying some older boys to the Charles River, and as we stood just beneath a bridge spanning the river, they proceeded to holler up toward its curvature—and magically to me, the sound of their voices came back in strangely garbled and drawn-out fashion. First time I'd ever heard an echo. My own voice wasn't as loud, so I found myself moving back and forth to get the maximum return for my own effort: Too far out, no echo; move closer, and I could detect an echo; closer still and I got my greatest payoff.

So when I first heard of bats using echolocation, I said "Sure, I know what *that's* like." All I had to do was imagine the bats having big-league skill at locating themselves that way in relation to the stuff around them, instead of being limited to my playground-level abilities: Send out high-pitched sounds, listen for the results, and make your move accordingly. In addition, I might add, when I later studied comparative neuroanatomy, I had no difficulty in observing the congruence between a bat's acoustic system and my own human variety.

Turns out, however, that Nagel was using the bat's echolocation abilities merely for illustrative purposes, wanting to describe as he put it *"a sensory apparatus so different from our own that the problem I want to pose is exceptionally vivid."[1]* The problem he was contemplating involved our difficulty in making a comprehensible jump from any organism's physical apparatus to the exact way in which it experiences things—and the related issue that in our effort to imagine what such experiences might be like for others, we have no more than our own subjective experiences to go on.

In the process, Nagel noted the ambiguity of our expression *"What it is like,"* emphasizing that he wasn't referring simply to what *it* may sort-of-resemble in our own experience, but *"how it is for the subject himself."* And that's in fact when difficulties start as I attempt to empathize with the bat's echolocation. For instance, I'm subject to an irresistible flow of inner words, and I can't seem to completely eradicate this *totally alien* experience (batfully speaking). And additionally, there's this sense **I** have of **my** self that's constantly involved in whatever **I'm** doing. I refer here to more than those occasional times when I'm explicitly thinking along lines like: "Now I'm doing this, and I guess I'll do that next" (what we might refer to as *introspective self-consciousness*). I'm referring rather to the sense-of-self that's constantly present in the background of my human experience, a sense that may well be absent in organisms with less complex neural circuitry.

PERCEPTIONS THAT I'M NOT UP TO EXPERIENCING

In any case, however, since the 1970s, scientists have continued to garner data on even more arcane sensory abilities of other animals, so I'm now able to replace the bat's echolocation with feats that I can't even get close to imagining. For instance, I can't come up with the foggiest notion of *what it's like* for the bluebirds sojourning on my homestead each summer when they start charting their course southward in autumn, *sensing* the earth's magnetic field as they go. I'm as totally unable to imagine that sensation as a congenitally blind human would be when it comes to imagining the bluebird's beautiful color. And even after scientists have completed a delineation of all the functional components involved in the magnetic-detection apparatus, I still won't have a clue about how migratory birds *experience* the earth's magnetic field—and that's one of the points Nagel was making.

WHERE DOES CONSCIOUSNESS STOP?

A feature of consciousness so obvious that we often don't pay enough attention to its likely significance is the fact that even the most perceptive people disagree as to when this property first makes its presence on the animal stage. Or as Nagel put it from the other direction: *"if one travels too far down the phylogenetic tree, people gradually shed their faith that there is experience there at all."*[2] The gradual shedding of belief in consciousness is of course just what we'd expect to happen if consciousness was indeed a basic attribute of nature—one that manifests itself with progressive forcefulness as behavioral complexity (and correlated computational complexity) increases.

CONSCIOUSNESS 'TESTS'

Problem is that we have no *objective* test for detecting the presence of consciousness, a property that of its nature seems ascertainable only *subjectively* (or as we sometimes say, from our inner experience). Added to that, our modern understanding of sensing systems, information-processing systems, and motor output programs, is such that we're *never* required to fall back on an assumption of consciousness in order to provide an explanatory account of an animal's behavior (Chapter 26). Hence, people usually end up drawing the line in somewhat arbitrary fashion, often at a point where they can no longer empathize with the creatures involved (Chapter 24).

There's a wide divergence of opinion as to when consciousness makes its appearance. Nagel, for instance, was confident that bats are conscious. For *"After all,"* he said, *"they are mammals..."* And it's true that the powerful mother-and-child bonding of mammals strikes a resonant chord of such intensity that it's difficult for us humans *not* to think of mammals as conscious. "Difficult," though not impossible, because as we've seen, Descartes and his followers denied consciousness to *all* the other animals. And even today, some theorists of a totally different bent (materialist rather than substance dualist) deny consciousness to most or all of the other animals. Why? Because these particular theorists believe that only our own beloved species has the complexity of equipment necessary to support conscious happenings.

THE TURING TEST

Though there's no objective test for consciousness—no mental Geiger counter available for discerning its presence and its intensity—scientists have tried to develop *operational tests* that would help guide and correct our immediate intuitions. Alan Turing, the great theorist of modern computers, described one such approach in his classic article, *Computing Machinery And Intelligence*.[3] His idea was to devise a computer of sufficient subtlety that one could not differentiate its verbal responses from those of a human being, even after extensive questioning. (Both the human subject and the computer were to be interrogated from an adjoining room, using typed questions and answers so that the computer would not have to look like a human or speak in a human voice.) While Turing did not maintain that his theoretical computer would actually *be* conscious, many of his enthusiasts judge that an apparatus functionally indistinguishable from us during our avowedly conscious activities should be granted the property of consciousness in the same spirit that we *"by polite convention"* grant to each other (the last quote being the humorous manner in which Turing made note of the fact that we don't even have a way of proving that our fellow humans are conscious).

THE MALEBRANCHE TEST

Coming at the consciousness-detection project from a somewhat different direction, I myself have labored mightily "to develop operational tests that would help guide and correct our immediate intuitions," and have finally hit upon a procedure that I offer free of charge to those current theorists who—usually while in the grip of their own conceptual models—maintain that household members like my cat 'Patchy' do not have the requisite information-processing equipment to support consciousness.

I call it the *Malebranche Test* in honor of that great Christian philosopher—he who was willing to put his foot where his theory was. Here's an account of the test's first recorded tryout:

> According to Trublet, Bernard le Bovier de Fontenelle once visited the Oratory on the rue Saint-Honoré and saw Malebranche kick a pregnant dog rolling at his feet. The dog let out a cry of pain and Fontanelle a cry of compassion, whereupon Malebranche said: "Well! Don't you know that it does not feel?[4]

My guess is that few of the current philosophers who limit consciousness to humankind would be able to carry their theoretical beliefs into their ordinary lives with such outstanding consistency (leastways, I hope they wouldn't). But when we're trying to set a lower limit to consciousness of a sort that we can humanly recognize, a useful question to ask ourselves would be: *"If I injure the organism, will it feel pain?"*

With that in mind, let's look at an earlier example: I mentioned in Chapter 24 that "One person I polled recently thought insects were not conscious; also that *frogs* and snakes were probably not conscious." But what if my friend had been asked to cut open a frog in order to observe, say, the peristaltic activity of its intestines? When these animals are actually used in biological experiments of this type, they are regularly 'pithed' beforehand. The procedure involves inserting a probe rapidly into the brain-case where the spinal cord exits, thereby instantly destroying the frog's brain. Why? Not just to prevent the frog from jumping around, because the animal is pinned down securely during experiments. The main reason has to do with a sense that the frog *does* have some sort of primitive consciousness, and more specifically, that it *is* able to experience pain. Bottom line: whenever we shrink from doing something lingeringly injurious to a particular type of organism—hence painful if that organism has a system for suitably processing such noxious stimuli—we are voting with our actions that this organism is probably conscious. (And I suspect, by the way, that my friend would've changed her mind about a frog's consciousness if she'd been confronted with an experimental need to inflict injury.)

CONSCIOUSNESS AND PLANTS

What about consciousness in plants? The question provides an opportunity to share with readers my enthusiasm for *Red Maples,* the predominant tree on my homestead. They're well named, because they provide different gifts of rubor throughout most of the year. During winter, for instance, their red buds against the whiteness of snow provide reassurance that Spring is not far away. And when that season finally does arrive, the ethereal red of their delicate efflorescence evokes fairy-tale fantasies. But for those who enjoy the excitement of fireworks, my Maples save their best for autumn, creating explosions of bright red, orange, and yellow, against a backdrop of many tall pines. In fact, now that I reconsider the matter, I guess Steven Weinberg was probably right after all: Nature does seem *"more beautiful than strictly necessary."*

In any event, since panpsychism posits a fundamental attribute of consciousness to all things in nature, the thesis would contend that my Maples are conscious too. The good news in this respect is that I've been known to talk to my trees—though the bad news is that they never talk back. Literally at least, although I think I am able to do a bunch of lip reading, so to speak, by observing their activities. We'll get to that shortly, but first a disclaimer that I'll illustrate by reference to a plant-tending acquaintance of mine. She speaks soothingly to all the plants in her home, and swears *that's* the reason they grow so luxuriantly. Seems they're comforted by her dulcet tones, which provide the key to their prosperity.

But since plants show little evidence of an acoustic system or of a computational system suited for decoding nonverbal aspects of the human voice, I've had to view her contention as another example of rampant animism. I did, however, proffer an alternate story to explain the vigor of her plants, namely that her solicitous attention to their placement and watering (along with a strategic use of 'Miracle Grow') accounted for their bountiful health. She concluded, in turn, that I was displaying one more instance of what she refers to as my deplorable cynicism.

Still, most folks would agree with me, because not too many people attribute *any* sort of consciousness to plants. The situation's strikingly different when it comes to animals though; and in this regard, it's worth noting a default program used by our information-processing systems for automatically distinguishing organisms we refer to as animals from those we call plants. The program applies a simple principle: *Animals move, plants don't.* And when I refer to this as an *automatic* program, what I mean is that we don't have to reason to the conclusion, but rather that we spontaneously experience things that way—which in turn sets us up for experiencing an illusion that we're about to come to.

In the meantime though, let's note that once we start reflecting on the movement/non-movement distinction, we have to make a significant qualification. That's because plants <u>do</u> move, albeit those in our ordinary midsize world do so mostly by growing (except for a few renegades like the marshland fly-catcher with its rapid closure of a specialized leaf). But the crucial point is that plant movement in the form of growth occurs *imperceptibly* from within the time-dimension of motion that our senses are able to register.

This sets us up for a strange illusion that may occur when time-lapse cinematography comes into play. For when the growth of a vine's branch, for instance, is speeded up so that we can actually *see* it moving briskly along, we're likely to have the same sort of automatic response that comes when we see animals in motion.

When that happens, we readily read motivation into movements: "Oh yeh," I might say in the case of a squirrel, "it's running up the tree to get away from Shep." And as I watch the vine's branch moving within time-lapse cinema, the same automatic tendency to provide a motivational story leaps into action: "Oh yeh," I want to say, "that branch is snaking up the trellis so it can take better advantage of the sun's energy."

It would be shear animism of course to actually attribute such articulate motivation to plants, because they don't seem to have the requisite systems up and running to underwrite such complex experience. Nevertheless, plants have much more information-processing power than we usually give them credit for—although as we're about to note, their systems work within a time-frame that makes it quite impossible for us to relate to these organisms by way of any consciousness that we might humanly imagine.

PLANTS AS 'MARTIANS'

Philosophical theorists-of-mind often emphasize that the higher animal nervous systems on our planet are, in principle, not the only possible vehicles for consciousness. Theorists have taken to using 'Martians' in a metaphorical sense to refer to *any* creature that would have the sort of complexly layered information-processing apparatus capable of underwriting consciousness, though constructed of other building blocks than nerve cells and their connecting cables. (Theorists are prone to figure that creatures with an alternately constructed apparatus of considerable subtlety would've had to develop on some other planet; hence, the term 'Martian'.)

But given the trick my nervous system plays on me by projecting motivated behavior spontaneously only onto activity happening at the speed of animal movements, and given our increased understanding of the intricate information-processing ability of plants, the more I get to wondering whether the theoretical quest for extraterrestrial 'Martians' doesn't bear a bit of comparison to Galahad limiting his mythical quest for the Holy Grail to distant lands—only to find that he could've come across this precious goblet right outside his own castle gate.

THE MULTI LEVEL INFORMATION-PROCESSING SYSTEMS OF PLANTS

Let's illustrate now the complex multi-level information processing of non-nervous-system organisms by observing the slow-motion behavior of one of my beloved *Red Maples*. But first, let's review some of the lower-level behavioral programs required by these trees:

They need to be able to detect crucial information about their environment; for instance, where they can find things they need, like usable energy and water. Then they have to be able to "go get it." Some aspects of the necessary programs are relatively hard-wired. To illustrate: Devices for taking water on board (roots) grow into the ground, which amounts to a highly adaptive feature, considering that the earth contains a more reliable source of water than the air above. By contrast, devices for tapping solar energy (leaves) get stacked on suitably placed shelves (branches) that grow out and up in all directions from the central trunk.

Rigid unfolding of these obligatory programs, however, wouldn't be adequate for the task at hand. In order to prosper, individual trees have to read their local conditions and alter their basic programs accordingly. So, for instance, roots will tend to develop more luxuriantly in those directions where water is more readily available, and branches will develop more effusively toward the sun. In order to read local conditions, sensory devices are necessary. Furthermore, adaptive movement in the form of growth requires computational programs capable of making sense of what's been detected and able to deliver coordinated instructions to the appropriate areas.

COMPLEX MULTILEVEL PROCESSING IN A NON-NERVOUS SYSTEM

The question naturally arises at this point: How is information to be conveyed to and from all the far flung reaches of the tree, and how is the information to be coordinated? After all, there's no principal command post equivalent to our centralized brain, and there's no web of nerve fibers to carry information rapidly to the most distant boundaries of the tree. So how do messages get conveyed? And where in the world do they come from?

Turns out that, like us, trees *do* have pathways over which information travels. And like us, they have chemical messengers to carry the news. Only thing is that while nerve trunks deliver information in a matter of milliseconds, the vascular channels of plants take their sweet time about it. That's because nerves convey information in a series of electrical waves traveling rapidly for sometimes a few feet at a time before releasing chemical messengers right next to their target. In sharp contrast, a plant's chemical messengers have to travel slowly along the entire route. (Animals also use this means of spreading the news, messenger-molecules coursing through the blood stream being referred to as *hormones,* a term that's also applied to the messenger-molecules used by plants.)

While that explains how messages are able to get back and forth from their various locations within a tree, it doesn't tell us where coordinated orders are coming from in the first place. And this question seems especially salient in light of our own subjective experience, in which we have the sense of a *unitary self* giving orders to our body, say, to stand up or sit down. Or this same *focal-point-self* may decide to rummage through its memory to find the name of the familiar-looking person who's approaching down the road.

But even in our own case, the retrieved information and the chosen responses require multiple interacting circuits that are distributed widely throughout the brain. That is to say, while certain functions require specific brain areas, all functions need broadly interacting nerve networks to produce their adaptive results. To illustrate, striate cortex in the very back of the human brain is necessary for any sort of conscious vision to occur, but what we normally 'see' involves objects that we come to recognize, emotions that accompany these objects, and a readiness then to approach or to avoid (or to simply discount the objects as not currently relevant). Multiple circuits from back to front swing into action simultaneously to produce this result, even though the experience itself is seamless.

Having all these brain circuits close together in one central locality enhances then the speed with which the already rapidly transmitting units can interact, but it's not as if a central processor is *ever* localized at one 'point'. For whether processors are bunched into quart sized skulls or stretched to the fullness of a tree, all processors are spread out in space. And the fact is that the distributed programs of trees can and *do* modulate results to fit the conditions at hand. For instance, limb growth on the particular Maple we're considering has accelerated toward the tree's sunny side and slowed on its shady side.

"That may be true," a critic might say, "but any candidate for phenomenal consciousness—in accord with modern theories—must at the very least demonstrate the presence of higher-level information-processing to monitor, supervise, and coordinate, lower-level programs." And as a matter of fact, trees easily meet this requirement, albeit in their own laid back way. To illustrate from the Maple in question: Its *branches-to-sunlight* program has resulted in fuller and wider branching on the meadow side. But if that particular lower-level program were to have had its way unchecked, disaster would have surely followed. Why? Because the tree would've become overweighted in the sunlight direction, and gravity would have eventually brought it down.

To prevent this disaster, the tree trunk has bent away from the meadow's sunlight, arching back toward the forest shade (as we might arch our own backs to compensate for a heavy bag of groceries we're carrying). In other words, this tree is exhibiting higher-level processing by altering the default pattern of its *trunk-grows-straight-up* program in order to keep its balance, while still permitting the *branches-to-sunlight* program sufficient leeway for the tree to harvest enough solar energy. And this same higher-level program-modulation has resulted in the growth of heftier roots (readily observable even at ground level) on the shady side, allowing firmer anchorage for the extra weight of sunny side branches. In short, this tree forms a case study in lower-level systems being exquisitely modulated by an immensely effective, though spread-out, higher-level apparatus.

PLANT CONSCIOUSNESS—ONE MORE TIME

Now, suppose we continue to explore the notion of consciousness as a basic attribute of nature by asking how we would envision its occurrence in the Maple Tree I've just described. Certainly the many detection and information-processing systems of this organism—subject as they are to the control of higher-level operations—involve a functional complexity light years ahead of the thermostat that we examined earlier. In fact, arguably, the tree's intricate and multi-level complexity makes it a candidate for the fulfillment of some of our current information-processing theories of consciousness (a 'Martian' right here on Planet Earth!). Nevertheless, our cognitive limitations make it totally impossible for us to imagine what such consciousness might be like.

In fact, one thing that jumps out at us when attempting to make the comparison is that our human stream-of-consciousness flows at a rate provided by closely clustered mechanisms transmitting messages at relatively high velocities. So even if some theoretically conceivable plant had all the information-processing capability of our own organism, the snail-like pace of its operation would be associated with a stream-of-consciousness s------o---------s------p------r------e------a------d---------o------u------t---------i------n---------t------i------m------e that we'd have no way of directly relating to it.

For practical human purposes then, it makes sense to say simply that plants aren't conscious, but with the unspoken qualification: *As far as any form of consciousness is concerned that we could relate to directly*. In short, although our panpsychic theory would attribute consciousness to plants, it'd be nothing like the consciousness that my acquaintance lovingly projects onto them: that they can hear her soothing voice, which in turn evokes a calming effect that allows them to prosper under her tutelage.

Having completed our brief assay of the human attempt to envisage consciousness in both upward and downward directions, we'll now proceed to further examination of the only kind of consciousness we know. This will also provide an opportunity to review our bedrock information-processing equipment with its obligatory use of *objects-and-their-movements.*

32

ME, ME, ME

I noted in Chapter 15, albeit a bit tongue in cheek, that there's little danger of our sense-of-fairness running roughshod, since our self-interest program is well poised to overwhelm it much of the time. And in fact, with that thought presently in mind, perhaps I should complete the earlier quote I borrowed from Voltaire. For after noting that *"Our benevolence toward our own species…is born with us and is always working within us…"*, he finished up by saying *"unless it conflicts with self-love, which must always prevail."*

This 'self' that we're so given to loving is the central object of our lives—so much so that it takes years for the average person to overcome an overweening interest in *"me, me, me."* I have some personal good news to report to readers in this respect, however, because I've been making some decent progress lately in overcoming an unbalanced interest in myself. Fact is, if I'm still hanging around for another few years of steady practice, I may manage to get sufficiently unstuck on me that I'll be really proud of myself.

At any rate, this self is not only the central object of our lives, it also comprises the pivotal instance of the way in which our information-processing apparatus works. From very early in life, that is, the core experience of a *self thing* further reinforces our acceptance of the other objects that we come across. Alternately stated, since I take the thing I call me totally for granted, I'm overwhelmingly disposed to accept the literal existence of the many other things that surround me throughout my life.

'ME' AS A CONSTRUCT

And given the fact that we experience our selves as not-the-same as our bodies, it's no wonder the thesis of substance dualism had such an enduring run within the human think-tank. Nevertheless, as we saw earlier, the concept of a separate substance to account for our sense-of-self could not withstand closer scrutiny

(Chapters 21-23). Hence, we need to view the self of our experience as a type of construct.

OUR "CENTER OF NARRATIVE GRAVITY"

Theorist of mind, Daniel Dennett, provided a neat analogy to clarify the notion that our "thinking thing" *(res cogitans)* need not be a thing in the literal sense. He did so by comparing our sense-of-self with a body's center-of-gravity, which after all is not an actual *thing* that's moving around inside—although it's certainly an important factor to take into account when dealing with our ordinary world.[1]

Whenever I want to remind myself of that fact, I think of the time I had my truck-bed piled high with stuff to help a friend who was moving. When I came off the Interstate at ordinary cloverleaf speed, my pickup's heightened center of gravity almost torqued me into a total tip-over. Real scary experience. But in my own defense, let me say that I've never careened around corners, especially since my 4x4 already has a high center of gravity for off-road clearance. Just that I didn't take its changed center of gravity sufficiently into account on this occasion, and almost paid dearly for my mistake.

At any rate, the present point is that our *center of narrative gravity* bears comparison with a body's *center of gravity*, in that both are very real in operational terms, and yet it's not as if in either case we could look within and identify some discrete object that's involved. In both cases, however, there *is* an ever shifting fulcrum that's always at dead center.

OUR SENSE-OF-A-FOCAL-SELF

To provide yourself with a graphic illustration right now of what's involved in this sense-of-self, try the following experiment: Look across the room at some object, say a chair. As you do so, you'll probably have a background sense that it's *you* who has just identified the chair. Now while still looking at the chair, shift your focus back to yourself, so that you become centrally aware that you're the one who's presently looking at the chair. I think you'll find that this will seem like the same you who was looking intently at the chair moments before, although now you are regarding primarily the you that's looking over there. At this point, shift your focus one more time, in order to view the you who was just regarding the you who was looking at the chair. This you will also seem like the same you who was regarding the you who was looking at you looking at the chair. (If *you* now try to regard the you who was regarding the you who was looking at the you who was looking at the chair, you'll probably start to lose track, because that's about as many 'yous' as you can hold together in the immediate-memory apparatus of *Homo Sapiens*.) But the point of the exercise is: However and wherever we shift our focus, we always seem to see things from the perspective of this same self—this same center of narrative gravity—even though there's no more of an actual *'thing'* involved than in the case of our body's ever shifting center of gravity.

CONSTRUCTING A MODEL OF THE 'SELF'

When it comes to constructing a conceptual model of this sense-of-self, I like to put the *feeling-of-intimate-familiarity* at its center, because this virtually omnipresent factor seems pivotal. We tend, however, to take this aspect of our human experience so much for granted that we seldom appreciate the fact that it's an add-on. Indeed, when it's claimed that our sense-of-self is constituted primarily by this relentless and utterly intimate experience of familiarity, common sense wants to reply: "It's just the other way around; my experiences feel familiar because they're _my_ experiences!"

Any person able to render such an analysis, however, is already a ripened human organism, that is, one who has achieved the developmental stage associated with a firm sense of self. And using this sense-of-self as a starting point, the most direct explanation does indeed seem to be: "Well, *of course* I have a sense of familiarity with this stuff; after all, *I'm* the one who recalls, say, how things were last night, and *I'm* the one who's seeing how things are this morning, and *I'm* the same me throughout, so *my* sense of familiarity naturally follows." In other words, whenever we use the 'self' as a take-off point, it's congenial to think that this 'entity' automatically imparts the sense of intimate familiarity to all our experiences.

But once we give up the homuncular view of a 'little man' at the control panel, this sequence gets turned on its head. For it's not then that some *'I' thing* brings about a sense of intimate familiarity. Rather, it's the relentless operation of our organism's apparatus for evoking a sense of intimate familiarity that results in our experience of personal sameness. (Indeed, even if there were some *'I' substance*, it would still require a mechanism, ghostly as that apparatus might be, to account for how the *experience* of personal ownership gets reliably fastened onto the *fact* of personal ownership.)

REVERSE ENGINEERING AND THE SENSE-OF-INTIMATE-FAMILIARITY

The easiest way to see that our sense-of-familiarity involves an additional operation is by "reverse engineering"—in effect, by noting those unusual instances in which the sense-of-familiarity either fails to kick in, or does so in inappropriate fashion. For warm-up purposes, let's look first at relevant mishaps of this sort that people have experienced in an outward direction: Thanks to Yogi Berra's famous malapropism, almost everyone's become aware of the odd state referred to clinically as *deja vu,* in which a person comes across something for the first time, yet an inappropriate sense of intense familiarity attaches itself to the experience. The opposite phenomenon, called *jamais vu,* also occurs, during which an individual loses the sense of familiarity in relation to something that he knows is actually quite familiar.

THEY PAVED PARADISE AND PUT UP A PARKING LOT

In one of Joni Mitchell's delightful songs, she provided apt lyrics to emphasize the point that we seldom miss what we have till it's gone—a fact that as just noted applies even to our *sense-of-intimate-familiarity.* And more dramatically still, this most closely held of our human underpinnings can also get thrown off kilter in an inward direction. Psychiatrists use terms like 'depersonalization' and 'feelings of unreality' to describe such events, during which a person may feel like an outside observer of her own mental and bodily processes—or in the most severe instances, may lose the sense of an *'I'* and a *'my'* almost totally.

Let me illustrate this sort of occurrence as it involved one of my own patients, a young schizophrenic woman who'd improved sufficiently that she'd been able to go home from the hospital. We met a short time later in my office, and while we were talking, she seemed in reasonably good control of her situation. Shortly after leaving, however, she returned in a state of agitated anguish.

When I asked her to tell me what was wrong, she said "I'm *losing* myself," a statement that was too vague for me to comprehend, so I struggled to get her meaning. She persisted in her effort to explain, and finally she was able to help me get the picture: What she'd been losing was her sense of herself. Not just the full sense of herself; _any_ sense at all of *her* kept slipping away. No wonder she felt terrified! For to permanently lose the sense of intimate familiarity that gets attached to all our experiences would be tantamount to the loss of our ongoing sense of self, which would be about as close to personal death as the enduring loss of our human consciousness that we mentioned earlier. Fortunately, this nightmarish lack of familiarity-feelings in my patient gradually subsided—thanks I think to our joint effort, the passage of some time, and appropriate medication—but my going through such an experience with her was enough to make me *not* take for granted the highly comforting sense of '<u>me</u>' and '<u>my</u>' that gets automatically and (almost) inevitably rubber-stamped onto our human experiences.

OUR STRING OF MEMORIES

The sense of 'my' also gets stamped onto the many memories that we resurrect throughout life, and this string of personal memories, like a necklace of gemstones, greatly enhances our sense-of-self. Still, we have only a modest number of memories overall, and the sense-of-self provided by a relentless experience of intimate familiarity may continue even when memories of one's personal past are unavailable.

To illustrate: While suffering the amnesia for one's past life that occurs during fugue states, a person loses the fully saturated sense of enduring through time, a sense that requires a backdrop of memory traces continuously intertwining with thoughts and perceptions of the moment. But even during such unusual episodes, the overall perspective from a *'hilltop of here'* is not lost, since the sense of intimate familiarity continues to operate. The afflicted person maintains *a* sense of self, albeit not the fully normal sense of self wherein the experience of 'I' is continually massaged by a flow of memories.

And in fact, if a continuous string of personal memories were absolutely essential for the continuance of our sense of self, then 'I' would have already died at least once. Let me explain: Some years ago while visiting my aged mother, I noticed a new picture on her favorite marble-topped side table—an old photograph of a little boy about two years old standing in front of an unfamiliar beach cottage. "Who's that?" I asked. She beamed happily (mothers are like that) and answered "That's *you,* dear." Seems she'd come across the fading photo while rummaging through some old stuff, and decided to have it framed for display. For a while after that, I tried to capture some memory from *within* this little tyke, the way I can muster up, say, a few memories of me "from the inside" during kindergarten at my old Agassiz grammar school. But nothing came. Nothing at all. And it occurred to me finally that in a way this little boy had already died, at least from a perspective of the sort of self that's available from strands of continuous personal memory.

FAMILIARITY/RECOGNITION

Disturbances in the ongoing sense-of-intimate-familiarity are often thought of as being associated with temporal lobe dysfunction. That's because certain temporal lobe abnormalities, such as those associated with partial complex seizures, are often accompanied by distorted experiences of this nature. It's important, however, not to limit our focus to some specific "brain center" that's to be judged solely responsible for the phenomenon.

So, for instance, when it comes to impaired recognition of one's own body (an experience that certainly involves a loss of the sense of intimate familiarity), disturbances of the right parietal lobe are quite commonly involved. Thus, individuals who suffer strokes in certain parts of this region may show no recognition of items that turn up in the left side of their visual field. Not only that, they will even tend to ignore the left side of their own bodies.

One such stroke patient whom I attended during my neurological rotation at the old Boston City Hospital years ago failed to shave the left side of his face, and never bothered to hook the stem of his eye-glasses over his left ear. Still more bizarre, if asked about his left arm, he'd deny ownership of it, even if you placed it in his right hand. And indeed, there'd been a similarly afflicted patient on the ward a while earlier who kept trying to heave his left leg over the side of his bed, insisting that the leg didn't belong there because it wasn't his own.

In summary, although we ordinarily take the sense of personal ownership of our selves and our bodies for granted, the feeling of intimate familiarity actually gets "stamped onto" all the contents of consciousness, providing us with a *'hilltop of here'* from which we survey our world.

MY ACTIONS

The sense of intimate familiarity also attaches itself to our actions, and helps to distinguish the normal experience of our intentional activities from the many complex crosscurrents of causes and effects that we see swirling around us. If I get jostled in a crowd of people, for example, I experience the movement as something that's happened to me within the ongoing rush of environmental events. But if I decide to push back, I don't experience my resulting movement as something that's just happened to me. In marked contrast, I experience the movement as *mine,* one that my focal self has initiated.

In a variety of neurological malfunctions, however, some of the movements produced by a person's voluntary nervous system will not be stamped with an active *'my',* leaving the patient with a sense of not being *'author'* to the actions. Anyone who's had a brisk knee-jerk response during a physical examination can get a benign sense of this sort of thing. Patients often look with amusement as their leg jumps, because they know that their own nervous system has indeed rolled out the action, but without their say-so. (In this example, the movement has been initiated at the spinal cord level, an area to which the *'my'* program does not extend.)

MY CONSCIOUSNESS "AT THIS POINT IN TIME"

One final comment about our human consciousness. The *present tense,* within our experience, is always spread out in time. For us to become aware of any experience requires a duration of many milliseconds. Flash a picture for, say, 15 milliseconds, and we won't be able to pick it up consciously—even though that's often enough time for a stimulus to be registered by our detection systems and processed in ways that become indirectly detectable (the sort of occurrences referred to as "subliminal stimulation").

To supply another example, good net-players in tennis will sometimes react to a shot canon-balled in their direction by hitting a "reflex volley" for a neat winner. Their response happens before they can become conscious of their response—hence before their sense of ownership can kick in. And this can leave them feeling almost like some tennis fan who's applauding the action from *outside,* as in "Wow, did I do that? Pretty darn good!"

In summary, human consciousness takes time to happen, and its occurrences are normally stamped with a *feeling of intimate familiarity* that adds an ongoing sense of *'my'* to whatever the contents of consciousness may be—which raises another significant question: Does consciousness always have to have one or another <u>content</u>? Academic theorists are almost at one in agreeing that it *does,* for how could a person ever simply *be* conscious without being conscious *of* something? In the coming chapter, however, we're about to question the validity of this consensus view.

33

"SOMETHING"

The event I'm about to describe happened within months of my graduating from Medical School. I was riding in what used to be called the automobile's "death seat," and the driver had, unaccountably, hit a roadside tree. My head smashed through the right front windshield, knocking me cold, and I came to a while later on the road (missed the now standard availability of seat belts by about a year). Once my head started to clear, it did so rapidly, and my very first act was to tell my toes to wiggle. I recall that vividly because of the intense feeling of relief I experienced when those extremities did what I directed them to do. (Having been exposed to the ravages of quadraplegia during a medical school clerkship at the West Roxbury VA Hospital, my first coherent thought was to check for a dreaded spinal cord injury.)

But the aspect of the situation we need to focus on now happened as consciousness was just starting to return. I can't describe anything before that, because during my knockout period, I experienced nothing. Total oblivion. *Nothingness.* It's important to emphasize that fact, because my reemergence was not at all like, say, waking from a sound sleep—or from general anesthesia for that matter—where there's a comforting sense of personal continuity. Instead, it was like coming back from the void. The events occurring early in this process are the ones I need to describe for our present purpose, because they pertain to the issue of whether consciousness, of its very nature, must always be blessed with one or another content.

Exactly how much time was involved in the transition from oblivion to alert consciousness I can't say, but the events that occurred involved oscillations of the following sort: first a *'something'* that was not the oblivion of total unconsciousness, then oblivion again followed by the *'something'*, then some vague flickerings of 'me', followed by the oblivion, and then the *'something'* once more—then some more vague flickerings of me, and back and forth. Then suddenly I regained full consciousness with a rush.

The *'something'* along the way is what requires explanation. It was a definite something, since it was <u>not</u> the nothingness of total oblivion. Yet there was no sense of 'I', and the *'something'* wasn't about anything; that is, it had no content. None whatsoever. Just that it was there, and that it was *not* the nothingness of total oblivion. Beyond that, there's little I can say about 'it'. In fact, anything I try to say tends to get off in the wrong direction. For instance, if I were to say *"This is an experience that I had,"* the statement would immediately mislead, because there was no sense at all of an 'I'—not the slightest shred of that background sense of self that manages to make it through even those tasks in which we become so involved that we seem to "forget all about ourselves."

Perhaps the best way then to express it is to say that I'd been in momentary touch with a "pure consciousness event," and that for some reason the event became recallable, rather than being relegated to the many many happenings of a lifetime that never become consolidated in our long-term memory.

"YOU CAN NEVER *JUST* BE CONSCIOUS"

Let's return now to my statement at the end of Chapter 32. I noted that academic theorists-of-mind seem to be at one in agreeing that we can't *just* be conscious. Consciousness, they're convinced, must always involve consciousness <u>of</u> one thing or another. Let's illustrate that view, as stated emphatically by neurophilosopher, Susan Greenfield: *"We are,"* she said, *"always aware of something at any one time. After all, it is a contradiction to be conscious of nothing."*[1]

Now, since a *"contradiction"* cannot actually exist in our world, how do academic theorists of mind deal with happenings of the above sort? Answer: What they seem, in fact, to do is to apply the Ostrich Principle, which may be formulated as follows: *Ignore inconvenient considerations for as long as you can!*

All folks who are fond of their own opinions find this principle highly useful. And the best part is, you can bring it into play without even being aware of doing so—which, by the way, is a downside for anyone like me who's fond of his own opinions, so I'm going to try not to leave out objections to the "pure consciousness event" that I've just described. In canvassing possible objections, however, I'm left a bit to my own devices, because virtually no academic theorist has addressed the issue (except in a tangential way that we'll get to later).[2]

THE "SOMETHING" POSES A CHALLENGE FOR OUR STANDARD MODEL

When it comes to ignoring the experience of *'something'* that I've referred to above, we should note that this finding does <u>not</u> fit in readily with the current *Standard Model* of consciousness. That's because the Standard Model rests so heavily upon the notion of complex multilevel information-processing, whereas the *'something'* seems to involve no information at all. It just *is*. So we have a case of proposed consciousness that seems to bear little relationship to information-processing, the heart and soul of our current Standard Model.

On the other hand, the *'something'* does fit in readily with a model that views consciousness as a fundamental property of all that is. Bear in mind, however, that the panpsychic model we're investigating does <u>not</u> discount the crucial importance of information-processing. As noted earlier, for example, destroy the speech systems within the human brain, as unfortunately happens in a common variety of stroke, and the victim will lose all consciousness of language. So information-processing systems are clearly <u>necessary</u> for our human forms of consciousness to occur, just that the operation of our complex information-processing apparatus is not <u>sufficient</u> of itself to account for consciousness—unless, as we saw, theorists are willing to hang their hats on a totally mysterious *"supervenience"* of consciousness within intricate multilevel information-processing systems (Chapter 25).

BUT IS THE *"SOMETHING"* AN INSTANCE OF CONSCIOUSNESS?

With that prelude, let's see how the *'something'* stands up to criticism. John Searle writes: *"Mental states only exist as subjective, first-person phenomena."3* And resonating with the introspective effort of the great David Hume, he adds: *"One can never just be conscious, rather when one is conscious, there must be an answer to the question, 'What is one conscious of?'"4* Well, if those are essential ingredients for consciousness, then by definition, the *'something'*—what I also referred to as a "pure consciousness event"—should not be categorized as conscious.

Fortunately, however, as theorist William Lycan has so aptly put it: *"None of us gets to kidnap words like 'conscious.'"5* Nonetheless, I think it's perfectly appropriate for a person to be called on to explain why he uses a word in nonstandard fashion. So let me now attempt such a justification.

First, let me emphasize that I agree fully with Searle's statement about consciousness—as it applies to our normal human consciousness. We've already seen, however, that although the *"first person"* aspect of our consciousness is virtually always present, exceptions do occur. That's because *"first person phenomena"* require the stamp of familiarity that in extreme cases falls away. The *'something'* experience noted above suggests in turn that while *"there must be an answer to the question, 'What is one conscious of?'"* within the sphere of our normal human consciousness, this aspect of our highly sophisticated consciousness may be an add-on also.

Yet why refer to the *'something'* as consciousness at all, even with the qualification "pure consciousness event"? Answer: Recall first of all our *Principle Of The Excluded Middle (either A or not A)* that we discussed in Chapter 9. Applying this principle to the present case, a particular event is either conscious *or* it is not conscious. My toe-wiggling decision, after I came to, clearly falls within the category of conscious. The oblivion of my knockout state clearly falls outside the category of conscious. But what about the *'something'*? Does the reader judge that it fits better into the category of <u>conscious</u>, or that it fits better into territory that falls outside that notion, namely the <u>not conscious</u>?

If a reader should ask "Why do I have to choose?", I'd agree that we should always maintain the theoretical option of remaining agnostic; albeit, we might seize the moment now to recall the ancient Greek story of "The Procrustean Bed" related in Chapter 12. To restate what we said after relating that venerable tale: *"Our verbal boxes are few, relatively speaking, but the real-world instances to be matched up with these conceptual containers provide unending variations around any given theme [and] we are called upon quite regularly to cut and twist the things we come across until they squeeze into one or another verbal container we happen to have available."*

Within the above context, I would suggest that the *'something'* fits more readily into the category of 'conscious' than being relegated to the oblivion of unconsciousness—from which state it was so readily distinguished. And even those who would choose to remain agnostic on the issue are implying at the very least that the category of 'conscious' is not an unreasonable selection. Another alternative of course would be to have our thingifying conceptual program come up with some intermediate item like 'semiconscious'—or perhaps 'protoconscious'? (More on this choice later.)

MAYBE HE JUST IMAGINED THE *'SOMETHING'*

A second objection might be that perhaps there never really was any *'something'* experience—that perhaps I fibbed about it, or that I just imagined it. Against the former concern, let me say at this very moment that "I cross my heart and hope to die" if I'm not being a straight shooter with my readers (enacting that formula gained one a certain amount of credibility as a kid growing up in my neighborhood). Other than providing that sort of assurance, or perhaps swearing on the…*uh*…Koran, I don't know what else I could say to convince you of my earnestness.

Regarding the second concern, namely that I might've just imagined this *'something',* let's return to the passage in Chapter 22 where I stated tongue in cheek that *"when I tell friends about the Leprechaun who frolics in my front meadow every morning, they don't seem to believe that the particular creature I'm telling them about actually exists."* In that section, I was noting that just because we have a word, we don't necessarily have an item 'out there' in the world that matches up with the word. But let's say that I'd been, in fact, recounting an actual hallucinatory experience of a Leprechaun. Although my belief about this wee creature's presence in the meadow would've been false, the fact of this mental event *as a mental event*—if I'd actually had such a hallucination—would still have been a fact. John Searle made this same point more abstractly in his book, *The Mystery Of Consciousness;* for as he put it, when it comes to conscious experience, *"the appearance is the reality."*[6] The *'something'* then is a mental reality that actually unfolded during my automobile accident.

ARE WE NOT RIGHT TO IGNORE A SINGLE ISOLATED REPORT?

Another concern flows from the fact that we're all experts, so to speak, about the spectrum of human conscious experience. We may be worlds apart in position and culture from, say, a Saddam Hussein or a Mahatma Gandhi. Nevertheless, we judge that we can at least comprehend the various elements comprising their experiences. In that context, recall my statement in Chapter 31 *"that I can't come up with the foggiest notion of what it's like for the bluebirds sojourning on my homestead each summer when they start charting their course southward in autumn, sensing the earth's magnetic field as they go. I'm as totally unable to imagine that sensation as a congenitally blind human would be when it comes to imagining the bluebird's beautiful color."* But suppose a fellow human assured me that he was privy to just such a sensation, and that he'd even undergo testing to demonstrate behaviorally his magnetic-field sensing ability. I'd think it more likely that he had some cleverly hidden magnetic device than that he actually had this type of sensation available to him. So it's easy for me to understand why folks would pretty much ignore my one little voice crying in the wilderness. Hence, we need at the very least to look for consensual validation concerning my report of "pure consciousness events."

SEARCHING FOR CONSENSUAL VALIDATION

The *'something'* experience of my accident did in fact cause me to keep my eyes open for similar reports. First description I chanced upon—one that seemed at least in the same ball park—occurred in William James' magnum opus, *The Principles Of Psychology*. James quoted Professor Herzen as saying:

> During the syncope [fainting spell] there is absolute annihilation, the absence of all consciousness; then at the beginning of coming to, one has at a certain moment a vague, limitless, infinite feeling—a sense of existence in general without the least trace of distinction between the me and the not-me.[7]

Since Herzen described the experience as being *"without the least trace of distinction between the me and the not-me,"* we were both on the same page in that respect. And I was happy enough with his mention of *"limitless,"* because any content delimits the whatever that's in thought or feeling from the rest of the stuff that's outside the given content. Hence, content is always limited, whereas the *'something'* I've been talking about was, in this specific sense, *"limitless."* I wasn't as keen about his reference to *"infinite feeling,"* first of all because the *'something'* was not a *"feeling"* in the common (emotion laden) use of this term, and secondly because *"infinite"* can so readily be given a grandiose spin that just wouldn't apply. I also had reservations about *"a sense of existence in general"* because that might imply the presence of a highly abstract concept (i.e. some content), although the professor probably meant no more than that there was *'something'* occurring, simply that it wasn't specified in any way.

I went through the above quibbling just now, by the way, mostly to make the point that when one is trying to describe something that lies outside our *'ordinary world'* by means of verbal tools that've been fashioned for dealing with our ordinary world, problems are unavoidable. In any case though, the good news is that I eventually came across a whole bunch of reports dealing with the "pure consciousness event"—a phrase I've been putting within quotation marks all along because that's what the experience is referred to in the available literature.

THE WORLD OF MYSTICISM

The bad news, however, is that most theorists-of-mind view the literature I eventually came across as irrelevant for their purposes. That's because by far the largest collections of such reports have been preserved by Institutional Religions—embedded within accounts of what are referred to as 'mystical experiences'. And since Christian mystics have rendered their accounts in terms of Union with God, Hindu mystics of monist persuasion in terms of oneness with the All, Sufi mystics in terms congruent with their Islamic world, and so on, such accounts have been deemed suitable only for perusal by partisans of a given religion. But this fact should not dissuade theorists of mind from examining the presently *relevant* phenomenon that's embedded within such culturally related presentations.

THE "PURE CONSCIOUSNESS EVENT"

Earlier, I said that theorists have ignored the occurrence of pure consciousness events "except in a tangential way that we'll get to later." What I was referring to is the fact that quite a number of academic theorists *have* reflected on such events, though not out of primary interest in implications for the theory-of-mind. Rather, these theorists have been concerned first and foremost with 'mystical experiences' themselves, of which the pure consciousness event is not even an always occurrent aspect. And in fact, when the event of pure consciousness *does* occur, it's almost inevitably surrounded by transitional states fore and aft during the meditation of adepts.

The eminent 20th century British philosopher, W.T. Stace, known best perhaps for his commentaries on Hegel, devoted an entire volume, *Mysticism And Philosophy*, to this subject, and he chose to give primacy in his analysis to the pure consciousness event, referring to it as the *"core mystical experience."* Here's his depiction:

> ...the Unitary consciousness, from which all the multiplicity of sensuous or conceptual or other empirical content has been excluded, so that there remains only a void and empty unity. This is the one, basic essential, nuclear characteristic...[8]

Congruent with this depiction, a more recent academic expert on mysticism, Robert Forman, defined pure consciousness events as *"experiences that are devoid of all content"9*—which of course is exactly the sort of experience to which I was referring. No expert in the area who's actually experienced a pure consciousness event (*e.g.* Forman himself) would even think to deny the existence of such occurrences. But given their rarity—at least the rarity with which such experiences get encoded in memory—the majority of academic experts end up analyzing a type of experience they've had no contact with themselves, hence one that's no more personally imaginable on their part than are visual phenomena to one who's congenitally blind. It's not surprising then that some of these theorists have denied the existence of any pure consciousness event, usually because they don't separate the event at issue from transitional experiences fore and aft.

To illustrate the point: It would be as if a theorist said to me: "How can you maintain that your experience while regaining consciousness after the automobile accident had no content? After all, you said that during this very episode you experienced 'vague flickerings of me', which certainly does involve content." I'd answer, of course, that the overall experience *did* indeed have content, but the particular event I've referred to as the *'something'* did not. In analogous fashion then, if a meditator describes an experience of, say, great joy, *that* part of the ongoing experience does have content (a feeling of bliss), but that's aside from the empirical issue of whether there was also a pure consciousness event—an experience devoid of all content—during another interval of the meditative sequence.

BIASED BY MY THEORY

I fear the present chapter has already gone on too long, yet there's one more concern that probably needs to be addressed. After all, some readers might well think it self-serving that I've lassoed such an extreme event as added support for my proposed theory of consciousness. So, as President Nixon might've put it, *let me say this about that:* "Yes indeed, I did go to considerable length in emphasizing the *'something'*, and I did so with malice aforethought because its occurrence does <u>not</u> square easily with the current Standard Model of consciousness, yet fits congenially with a neo panpsychic model."

However, I think the major impact of the actual event was on my own development as a theorist, rather than as merely one more straw that I'm trying to throw on top of the persuasivity pile. Having immediate access to such an event—one that was totally outside our ordinary human consciousness—provided me with an insight that I might otherwise never have come by (at least not in such vivid fashion); namely, that our human consciousness is not THE consciousness, but rather a particularly sophisticated type of consciousness that's been programmed within our own grand and glorious species.

And I would go so far as to speculate that if the theorists-of-mind mentioned in this book had been fortunate enough (or unfortunate enough, given the circumstances) to have gone through this same sort of experience, they would not have moved so quickly to identify consciousness *per se* with crucial aspects of our own human variety—a kind that's always experienced subjectively, and that always has content (not to mention an inexorable kind of content, namely *objects-and-their-movements*).

In the upcoming chapter, we'll draw out further implications of the *'something'* for understanding our normal human experience—a project that I hope will be of special interest to any reader who's ever been fascinated by the sometimes highly publicized reports of "mystical experiences."

34

THE "SOMETHING' EVENT AND OUR HUMAN WORLD

In the last Chapter, while suggesting that the *'something'* does indeed deserve to be classed within the category of 'conscious', I made note of the fact that we can always create intermediate categories, and offered the term *"protoconscious?"*. Reason for the question mark was that I don't really consider my exposure to the pure consciousness event as an instance of 'protoconsciousness'. Here's why:

The events that occurred while I was coming to after my trauma involved a back and forth among the *"oblivion,"* the *'something'*, and *"vague flickerings of me."* Such oscillations require time, and that fact suggested a well-known aspect of human consciousness. For as noted in Chapter 32, our consciousness requires nervous-system elaboration over time *("The present tense, within our experience, is always spread out")*. This durational factor has caused me to speculate that some apparatus in my brain had initially rebooted, a system needed for basic consciousness to be sustained in a way that becomes both experienceable and recoverable in human memory. I then envisioned such a *sustaining device* as composed of neuronal circuits that in the instance of my accident had gotten back into gear first.

Such an explanatory story would be congruent with our knowledge that neural circuits like the *reticular activating system* down in the the brain stem must be up and running in order to produce the generalized activation associated with consciousness. This thesis would also explain the difference between a *'something'* experience and the oblivion. Presumably, while oblivion prevailed, nerve cells within these basic circuits were temporarily silent, or firing randomly rather than in a normally patterned manner.

CORRELATION, ONE MORE TIME

Since explanatory stories are judged nowadays, quite rightly, on whether they can be operationalized into testable procedures, let me note that relevant experimentation is possible, at least in principle.[1] For instance, acute trauma patients in states of relatively short term unconsciousness could have their brains imaged sequentially (c.f. Chapter 24) during their obtundation, and then debriefed <u>immediately</u> on regaining consciousness. This procedure would permit at least rough correlation between types of brain activation on the one hand, and reports of the *oblivion,* the *'something',* and *flickerings-of-self* on the other.

I include the *'something'* event because, within the experimental setup just described, I think such experiences would turn out to be not at all as rare as they presently seem to be. That's because most people who've been knocked out seem to have better things to do with their time than to make careful note afterwards of *ephemeral-experiences-without-content,* so such events don't get sustained in memory. (Markedly different of course is the situation involving those who manage to achieve states of contentless consciousness while involved in advanced levels of meditation, because such adepts are highly motivated to retain the fruit of their practice in memory.)

Let me add that even with immediate debriefing, I'd expect only a small portion of such accident victims to report *'something'* events. That's because, in the majority of cases, the sustaining circuits I've hypothesized <u>and</u> our information-processing systems would likely reboot at about the same time, so the initial return of consciousness would have content. My proposal implies an occasionally desynchronized return-to-action of ordinarily well coordinated systems.

In order to clarify this point, let's take a look at another case of smoothly meshing systems within the human brain that on occasion become uncoordinated: During sleep dreams, our bodies don't move because the nerve pathways triggering our muscular apparatus have been temporarily decommissioned. That's kind of a neat feature, since if we're running and jumping within a dream sequence, the fantasized actions don't get translated into inappropriate movements that might injure us while we sleep. As soon as we awaken from a dream state though, the nerve pathways that stimulate our muscles become unblocked. But not always! On exceptional occasions, waking consciousness returns first, and the rebooting of movement pathways lags behind, creating a short interval of so-called "sleep paralysis," wherein we're awake, but can't yet move.[2] Analogously then, I'm hypothesizing a temporary desynchronization of our complex consciousness apparatus to account for human contact with pure consciousness events.

ME AS MYSTIC?

If, as Stace suggested, the *"core mystical experience"* involves the pure consciousness event, and given the fact that I've had contact with pure consciousness at the time of my accident (and on some subsequent occasions), does that make me a mystic? If so, then I ought to be pleased as punch, because contact with such events has been said to lead to a state of *"enlightenment"*—the term used in mystical literature—and I didn't have to devote years of apprenticeship in ashram or monastery!

Let's look for a moment now at what's meant in a mystical context by *"enlightenment."* Obviously, it can't refer to enlightenment about things that interest us greatly in our ordinary world; for if one reads accounts of, say, the workings of our human body as portrayed in some of the venerable writings of adepts, what jumps out is the presence of conspicuous <u>un</u>enlightenment. So 'enlightenment', if the term's to be justified, can't refer to specific knowledge of things in our ordinary world.

As might be expected then, the term 'enlightenment' was applied to a state of radiance involving what lies *outside* our ordinary human world.[3] And pure consciousness events were particularly fruitful when it came to achieving a sense that our world, though managing to keep us vividly in its thrall, is a world of illusion (*'maya'*). Eastern mystics, especially, emphasized that the world of our senses is illusory; and more crucially, that even the beloved 'self' of our everyday experience is illusory. Notwithstanding, as they viewed it, we tend to remain strongly chained to our human illusions, and our associated *"attachments"* are responsible for all the anguish we experience throughout life. In sum, their enlightenment was *"not of this world,"* so to speak, though having important implications for the project of how best to deal with our ordinary world.

MY ENLIGHTENMENT?

A reader conversant with mysticism might wonder at this point whether my contact with pure consciousness led to my own enlightenment, as has happened with mystics. Well, I would claim that exposure to pure consciousness events has promoted an enlightenment of sorts, if only by way of developing vivid appreciation for the fact that within my human states of consciousness, I'm inexorably limited to the way my own species' cognitive apparatus works—including importantly my compulsory reliance on the construction of *objects,* both physical and conceptual.

And after all, even the brilliant Spinoza could do no better than resort to our same human apparatus, as for example when he spoke of *"ideas"* in God. For though he made repeated efforts to distinguish between human ideas in God and the *"eternal and infinite <u>intellect</u> of God,"* I don't think he managed to avoid indulgence in yet another instance of our anthropomorphic projection. For *"ideas"* and *"intellect"* refer to human stuff. It would hardly fit with the infinite capacity of one necessary Substance to be limited to clumsy instruments like our conceptual containers—wondrous as words/ideas have been for humankind when dealing with our ordinary world.

Back to the issue of my…*uh*…enlightenment, the vivid recognition of 'my' self as a construct does modulate my powerful gut-sense of a self that's utterly basic. In fact, it's become automatic now for me to recognize that my own 'self' is an invincible illusion experienced within the ordinary workings of my human organism. And it seems equally natural now for me to view the renderings of 'my' senses as illusory also—especially given the fact that this judgment is so strongly buttressed by the evidence of modern neuroscience.

IF EVERYTHING'S AN ILLUSION, THEN WHY BOTHER?

I expect, however, that readers with common sense may take a dim view of such an enlightenment. For if everything's mere illusion, then why bother? With anything? And as a matter of fact, teachings of the Eastern sages have often been rejected for just this reason. They've been accused, that is, of offering *'Nihilism'* under the guise of *Wisdom*.

It's difficult, however, to reach the heart of what the gurus were getting at just from reading some of their stuff—or more often, from reading what devoted enthusiasts think was meant by what a given guru was said to have said. We could look, for instance, at Gautama Buddha's famous dictum that our human *"attachments"* lead to all our humanly experienced anguish. A corollary of this teaching is that a state of total *"detachment"* would release us from all our anguish. That being the case, some of the Buddha's current enthusiasts tell me that the great man was holding up perfect detachment as the highest ideal in life. (And of course, recognition that all the stuff around us amounts to no more than *"a paper moon floating over a muslin sky"* might make it easier for us to work toward this wondrous ideal.)

Problem is, if we ever managed to achieve such an ideal, we'd also be *"released"* from anything we find humanly meaningful. And considering that each of us votes best by his actions, it's hard for me to think the Buddha actually embraced total detachment as his ideal. For he parted with the ascetic's life of *detachment* from the pleasures of this world—with its implied rejection of our human world as not sufficiently worthy—and he showed a good deal of interest in helping others, demonstrating thereby a vigorously altruistic *attachment* to his fellow man.

Now, at this point, one or another devotee might say: "The way you used the words *'attachment'* and *'detachment'* just now distort their actual import. What the Buddha was *really* getting at was [fill in the devotees's particular opinion]." And for all I know, their interpretations may be far more evocative of the great man's actual intent. Indeed, the problem of trying to figure out *exactly* what another person *actually* meant has been just one more problem over the years for Guardians of the conviction that we can indeed know what really is. But maybe we could use the present instance to review an earlier point about words being like boxes (Chapter 12). We noted then that ideally the content of each word *"should come in a standard-size box, but the box turns out to be so variable in size that more items or fewer items get packed into recurrent orders. And worse still, not only is the box likely to be of different size the next time we order it up, some items we thought we were ordering might not even get packed inside the next time around."*

In any case, however, I like to think that the Buddha was stressing primarily the need for properly modulating our human attachments. Because if he were totally disagreeing with the ole romantic notion that *"It's better to have loved and lost than never to have loved at all,"* my personal enthusiasm for his teaching would begin to wane. I might even find myself…*uh*…retreating to Aristotle's position that any virtue becomes a vice by excess or deficiency. That notion, of course, would apply also to the virtue of *"detachment"* from our human illusions, especially given the way I view our innate illusions.

HUMANLY INNATE ILLUSIONS AS APPROXIMATIONS

The findings of modern neuroscience certainly support the judgment that our representations of the world amount to illusions—in the sense, however, that our human representations provide no more than highly simplified approximations of what is. But such illusions are not *mere* illusions, because they provide us with the workable accounts needed for successful adaptation to our ordinary human world.

And why should we turn our backs on this world? Despite the many occurrences that we experience as utterly horrific, our world offers, on balance, such a beautiful, engaging, and enchanting view of *'What Is'* that our brief moment of human existence seems worth savoring to the fullest. Doing so, in turn, involves acceptance of our sensory pleasures, in harmony with the social pleasures we experience with others of our kind, in harmony with the esthetic pleasures we take in the beauty of nature and our own artistic creations, and in harmony also with the intellectual pleasures we have available within our human mode of thought.

Yet there seems little question that we often do take life *too* seriously, especially when we equate the successful execution of a certain path in life as "make or break." So often, for instance, I've seen folks eaten up by their own sense of failure in life, often because they simply didn't heed the notion that one's best is always good enough.

Vivid recognition that the world of our experience is an illusion of our transient selves does *not* in any case remove us from our own unique window on 'What Is'. And as long as we're talking about *any* member of our species, that member will always experience things in our own species-specific manner—give or take a few events that fall outside of our organism's ordinary workings. But realization that our human world involves us in a running series of illusions (including the core illusion of our 'selves' as discrete and enduring substances) does offer an opportunity to leaven our tendency to take ourselves and our lives *too* seriously. Or as a young nun so aptly rhymed the point over fifty years ago:

> Life is not a complex thing
> as many men have said;
> it's just an interesting game
> with all the winners dead.

These were words spoken by a youthful idealist, not by an old cynic. The difference, as is so often the case with words, lies in the context of the user. At any rate, there's something to be said for a balanced detachment that comes from not taking things *too* seriously. And sometimes I like to think that's what the man known as Gautama Buddha had in mind.

BACK TO OUR ORDINARY WORLD

While exposure to pure consciousness events can be helpful in an especially vivid way, experiences of this nature are surely not necessary in order to achieve worthwhile enlightenment. (Nor would their exposition be absolutely necessary for the purpose of warranting a modern panpsychic theory, though the fact of their occurrence is by no means irrelevant to the task.)

Thus Spinoza certainly appreciated the magnificence of 'What Is' to the point of becoming *"God intoxicated,"* though lacking any contact with pure consciousness events. And as a matter of fact, it's hard for me to believe that a fellow like Stephen Weinberg, given his articulate insights into the nature of our world, doesn't experience similar feelings of reverence and awe in the face of Spinoza's God. My guess is that Weinberg's no-nonsense way of downgrading *Deus sive Natura* in favor of more antiseptic notions like *"order"* and *"harmony"* may come from a sense that grown men don't cry—even when they're crying from joy. At least not in public.

ASIDE TO ENTHUSIASTS OF THE MYSTICAL TRADITION

Enthusiasts of mysticism might conclude that I've been downgrading the help provided by writings of the mystics with whom they resonate. "Where," they might ask, "is the sense of sublimity in your report? And where is the blissful union that's involved in *genuine* mystical experiences? Like a typical scientific nerd, you've offered little more than the wash water of mechanistic musings, while discarding all that's truly uplifting and divine."

In response, recall that I was limiting my analysis to the pure consciousness event, and as I noted: *"the pure consciousness event is not even an always occurrent aspect [of mystical experiences]."* In fact, even when mystics make an effort to describe episodes of contentless consciousness (a task that's simply not possible in any literal sense), transitional events usually take up far more of their descriptive time. That's because the transitional experiences usually occupy a far greater amount of meditational time, and, even more importantly, these experiences <u>do</u> have readily describable content (*"if a meditator describes an experience of, say, great joy, that part of the ongoing experience does have content..."*) In profound contrast, there's no content whatsoever to describe in a contentless experience. The significance of the experience to mystics then comes from the very fact that one's sense-of-self is absent, as is the this-or-that content of our bite-by-bite limited human consciousness. Thus, exposure to the event brings one outside our otherwise all-prevailing human 'dream'.

Mystics invested in a particular culture quite naturally interpret the loss of their 'self' in a way that's congruent with their own highly valued culture. Feelings of ecstasy and bliss that so often occur during the transitional periods of their meditation convey for them an apt response to oneness with that which is beyond our limited human nature. Congruently then, *"Christian mystics have rendered their accounts in terms of Union with God, Hindu mystics of monist persuasion in terms of oneness with the All, Sufi mystics in terms congruent with their Islamic world, and so on..."* (Chapter 33). Hardly surprising, after all, that the individuals involved interpret their overall experience within the coin of their own cultural realm. It shouldn't be surprising then that I do likewise; specifically, that I interpret contentless experiences from within my own culture of science, using a naturalized underpinning provided by *Deus sive Natura*.

Have I experienced periods of bliss, serenity, and awe? Yes, as a matter of fact, but such experiences don't hang on the *'something'*—and obviously they're not central to my project of the moment, developing a theory of consciousness. Nevertheless, let me note the fact that a more all-pervasive background sense of joy and wonder comes from the daily privilege of experiencing *'What Is'* in our inimitable human fashion. That's in fact why I prefer Spinoza's reverential phrase, *"Deus sive Natura,"* to more workaday words like *"order"* and *"harmony."* And as Spinoza, that great poet of natural philosophy, put it, *"The more we understand particular things, the more do we understand God."*[4] Of course, other folks might prefer such a notion expressed in the emotionally powerful lines of William Blake:

> To see a world in a grain of Sand
> And a heaven in a wild flower,
> Hold infinity in the palm of your hand,
> And eternity in an hour

After all's said and done, however, perhaps half of those with a keen interest in understanding how our species represents the world do not adhere to a fully Naturalist account (as in *Deus sive Natura*) because their enumeration of *'What Is'* includes both our world <u>and</u> an outside cause thereof. Their reflections, that is, occur within a conceptual framework involving **1.** Nature and **2.** a Transcendent Being who has created Nature. In the final chapter, we'll assess how the present model might resonate with their Supernaturalist world view.

35

ETERNAL LIFE AND PERFECT JUSTICE

There's an oft told story about a meeting between the great scientist, Pierre Laplace, and Napoleon during his heyday. Seems the Emperor was impressed with Laplace's Naturalist view of the world, yet puzzled about where the notion of God fit into such a universe. When Napoleon raised the issue with Laplace, he answered: *"I have no need for that hypothesis."* Since the present account, in its own way, also stays within the bounds of nature—nowhere requiring intervention by a *Transcendent Being* from outside of nature—the same question might be raised about the *World According to Homo Sapiens*. With that issue in mind, I should mention that my lack of need for the hypothesis does *not* mean that an alternate explanatory story for our world (based on the existence of a Transcendent Being) logically invalidates itself. So before ending, let me address that point.

NATURAL SELECTION OF CONSCIOUSNESS

First, however, I'd like to acknowledge my great debt to the elegant *Theory Of Natural Selection,* which not only provides an explanation for the development of a species' physical characteristics but also its behavioral programs. And not just basic programs allowing animals, say, to move their bodies in coordinated fashion. Social animals have built-in programs that promote cooperative behavior. And at the very top of the social-complexity pyramid, our own grand and glorious species has a particularly refined program of that nature, one that we experience consciously as our human "sense of fairness."

Nevertheless, the Theory Of Natural Selection creates its own problems when applied specifically to consciousness. For in days of yore, "conscious choice" would've been the automatic explanation given for the behaviors chosen by higher animals. But nowadays, their choices are readily understood as arising from relevant information-processing systems whose computations have been adaptively weighted (c.f. Introduction to Part II). And within this context, the addition of a separate conscious causality has become almost a pariah, seeming at times to do little more than create the 'causal closure' problems discussed in Chapter 26.

Especially is this the case for those who espouse the current *Standard Model* of consciousness, given their concomitant reliance on Natural Selection. In accordance with the latter theory, only traits that are useful in themselves get selected, and this fact serves to highlight the 'causal closure' issue. That's because consciousness must do something additional, of itself, in order to explain why *it* would be naturally selected. Yet, if consciousness really *does* cause something additional, then the 'causal closure' problem becomes even more acute. That's because the 'whatever additional' that consciousness does either is based on the normal functioning of the underlying neural networks,[1] or it is not based on their activity. But if it *is* dependent on the workings of relevant neural networks, then the same workings of those same networks should produce the same behavioral outcome without having to appeal to consciousness. On the other hand, if consciousness produces something additional that's *not* based on the workings of the underlying neural networks, we'd be back to a spooky causal agent operating independently of its neurological base.

Happily, the notion of consciousness as a basic attribute of our universe provides ready solution to this problem. That's because, if basic consciousness is *always* present, then its occurrence does *not* depend on Natural Selection, but rather only particular *kinds* of consciousness that are proportionate to related neural mechanisms (e.g. systems associated with language/concepts being naturally selected in the primate species known as *Homo Sapiens).*

APPROXIMATIONS—ONE MORE TIME

There's another implication of Natural Selection, however, that's more directly relevant to the present chapter. We've noted from this book's beginning that our cognitive systems have been selected for their ability to provide the relatively rapid and dependable approximations needed for adaptation to our *ordinary world*—rather than, for example, their ability to achieve the truth, the whole truth, and nothing but the truth. Focusing on that fact, however, a thoughtfully Religious reader might ask: "If, as you maintain, all we have are fairly rough approximations of what is, how come you're so set on ignoring, if not outright denying, the God of Judeo-Christian tradition? You seem, that is, to have no qualms about promoting your own view as the correct one, as if 'Guardians of the conviction that we can indeed know what really is' may not know, but you sure do!"

In response, let me say that I'm not trying to be dogmatic. Insofar as what I've said makes sense to readers, I'd be wildly enthusiastic about their incorporating such thoughts into their own world view. And in so far as some of the things I've said don't make sense to them, I'm happy to have readers leave my suggestions behind. (In fact, I feel totally honored that a reader has spent this much time attending to what I've had to say in the first place.)

From an expository perspective though, it's often useful to state one's opinions affirmatively, leaving out too many *"ifs, ands, or buts"* along the way. And I guess that's why I've acted a bit like the ole emergency ward physician in Chapter 18, who admitted that he didn't always know what he was doing, but that he was never in doubt. Of course, that sort of stance can amount to a considerable downside for anyone who's fond of his own opinions. Like me. So let me affirm explicitly now that I've been simply trying to provide in Part III a plausible explanatory story concerning the central enigma of consciousness.

THE SCIENTIFIC WORLD VIEW

Let me also use this occasion, however, to extol one of the virtues of a scientific world view: Since scientists make no pretense of attaining "the truth, the whole truth, and nothing but the truth"—thinking only in terms of achieving ever more useful approximations—they and their fellow travelers tend to be much more tolerant of divergent views. Take me, for instance. I'm totally tolerant of all the stupid jerks who call themselves theorists-of-mind, yet who reject my panpsychic model of consciousness. Well...*uh*...at least I'm not in favor of rolling out thumbscrew and wrack in order to help them acknowledge the error of their ways.

EFFECTIVENESS OF THE NATURALIST WORLD VIEW

An additional upside of the scientific world view has been its cumulative explanatory effectiveness over the years. In marked contrast, Supernaturalists have jumped time and time again to shortcut explanations for this or that phenomenon, looking to the "Author of Nature" for a direct and immediate explanation of whatever's involved, only to find their supernatural explanations eventually displaced by natural ones. This has given rise to the rather pejorative title, *"God of the gaps,"* to any Deity who gets rolled on line for direct causal interventions, but whose hands-on effort can later be dismissed as unnecessary.

SCIENTIFIC EXPLANATION IS NOT EGALITARIAN

But scientific tolerance is not to be confused with the sort of egalitarianism that would declare all views equal. Scientists after all are expected to follow certain rules (including obvious ones like *"it's a no-no to fudge experimental data"*). Also, in order to be taken seriously, conclusions have to be congruent with the logical principles built into our human information-processing systems. This includes the need for appropriate consistency.

Let's kill two birds with one stone now by illustrating this point while at the same time reviewing an earlier one. In Chapter 26, we noted the problem of 'causal closure' that tends to arise when we provide a brain-based causal account of some action <u>and</u> a conscious causal account for the same action. As we saw, once nervous-system activity is accepted as the cause of a person jumping back onto the sidewalk to avoid a careening car, then the explanation based on consciousness as an *additional* cause becomes superfluous—unless, that is, one retreats to a two-substance theory. Then the brain is viewed as perceiving the hurtling object and conveying this information to the consciousness substance, which then makes a decision and quickly tells the brain to move the body out of harm's way. By thus turning the situation into a *causal sequence* involving interaction between two different things, body and mind, the *causal closure* problem is averted. But scientific theorists who've previously rejected Descartes' theory of two separate substances for other reasons can't sneak it back in here just for their *ad hoc* convenience. To do so would be logically inconsistent, hence unacceptable.

A TRANSCENDENT BEING *AND* CREATION

How does the notion of logical consistency apply then to explanations based on two categories of substance: A Transcendent Substance that necessarily exists, and all the substances (things) that this Substance has created? Well, if it's logically inconsistent to posit more than one substance, as Spinoza judged it to be, then this traditional approach, using the notion of God *and* His creation, would have to be discarded by any straight thinker.

Only problem is that while Spinoza's thesis provides a useful model of 'What Is', I don't think his argument is airtight. For example, Scholastics never used the term 'substance' in only one way (*'univocally'* as they would've put it). They were quite aware of a distinction between the One Necessary Substance that had *within itself* the reason for its own existence and what we might refer to as subsidiary substances which, while having sufficient independence to perform John Locke's prescribed activity of containing the properties manifested, still depended on the One Necessary Substance for their moment of existence. (A modern way of making the same distinction would be to speak of 'substance' in the strong sense and the weak sense.)

While attempting to describe this fundamental Being—the Ground of all that is—within the awkward conceptual containers available to our species, Scholastics spoke of this Being as both "Transcendent" (above and beyond nature) and *"Immanent."* The latter aspect was expressed in the old Baltimore Catechism for children by saying: *"God is everywhere."* I'm not sure that the fancier expressions of Scholastics do much better, but the notion involved a *support from within* of everything that exists. And once one homes in on this notion (fuzzy as it remains because it's not within the direct purview of our ordinary world), the distinction between an *Immanent-Being-and-His-creation* on the one hand, and *One-Substance-And-Its-Modes* on the other, may become a bit too precious. At any rate, I don't think an alternate approximation expressed by the notion of "God and His creation" has been shown to be logically inconsistent, so I'm not about to argue that reason *requires* those holding such a view to mend their ways and relinquish their hypothesis.

What consistent reasoning certainly <u>does</u> require, however, is the rejection of a humanoid god—for instance, the God of Abraham, Isaac, and Jacob, as *literally* portrayed in the Bible stories of Judeo-Christian tradition. And it's this Superman God "whose eye is on the sparrow" that seems to actually interest the majority of believers.

"DO NOT FORSAKE ME OH MY DARLING..."

So how might we argue in favor of staying with the historical notion of a Transcendent God? One way would be to start with the fact just reviewed that our knowledge involves no more than fairly rough approximations; and since these approximations are directed primarily toward adaptation, there's lots to be said

for <u>choosing those approximations most helpful for the conduct of our life-strug-</u>
<u>gles</u>. In that regard, the notion of a Transcendent God seems to help many peo-
ple when dealing with the visceral angst evoked by the operation of our most
deeply embedded programs.

As a way of illustrating this point, we might start with the theme song from an
old Western movie, *High Noon*. Seems the hero felt compelled to risk getting
himself shot to death by a nefarious villain and his cronies. Yet according to Tex
Ritter's deep voiced rendition of the lyrics, the hero claimed *"I'm not afraid of*
death, but oh, what if I lose my darling…" (that would be his bride, whose Quaker
nonviolence could only countenance hooking up with him if he'd relinquish his
sheriff's work in the gun-battling ole West). But whenever I heard the above line,
I couldn't help think of it as a banner instance of "whistling in the dark." That's
because we're so strongly programmed to long for more life and to fear death that
I just didn't think either Tex Ritter, Gary Cooper, or the hero he was portraying
(or you or me) lack a robustly built-in *fear of death*. And after all, since the hero
feared losing his loved one, he'd have to figure that if he died, it'd be just one
more way of losing her—in addition to everything else he loved in life, his own
self included.

Now, if we can provide ourselves with a coherent account of 'what is' that permits us to avoid our own deaths (at least ultimately) and that of our loved ones,
why not choose *this* approximation over a less practical one that offers no hope of
continuing our 'selves' after death?

SLAKING OUR THIRST FOR FAIRNESS

And another of our core programs is the socially and personally crucial sense-of-
fairness program. I say "socially <u>and</u> personally" because, while we feel we should
treat others fairly, nevertheless our sense-of-fairness program normally peaks in
intensity when it comes to our own selves. We thirst for fairness in a world that
just isn't fair. But if we can provide ourselves with a coherent account of 'what is'
that permits us to believe we will eventually be treated fairly, why not choose *this*
approximation over a less emotionally satisfying one that offers little hope of ultimate fairness within the world of our ordinary experience?

INGREDIENTS FOR A SUCCESSFUL INSTITUTIONAL RELIGION

Examination of successful institutional religions shows that in fact they all provide supportive answers to these core concerns for *more life and more fairness*. The Christian religion of my birth, for instance, provided a supremely happy and trouble-free life in Heaven for the good guys, as well as punishment in Hell for the bad guys—many of whom seem to live high off the hog for their entire earthly lives, so they sure deserve their eventual comeuppance. Interestingly, the Jewish forbears of Christianity may've been rather late in offering the hope of a personal afterlife, but it was present in spades by the time Spinoza was drummed out of his 17th Century Amsterdam community (his disavowal of personal immortality seems to have been one of the key factors in his excommunication).[2]

What about Buddhism and monistic Hinduism though? Don't they preach total loss of the 'self' in Nirvana rather than life after death? Well, yes and no. These Religions are big on the notion of *reincarnation*. That is to say, ordinary folk are not *really* going to die, because they're going to be reborn again, and again, and again, until they've achieved a state of enlightenment that allows them to be freed from the wheel of rebirth. So the ordinary person who's chained to our human illusions can look forward to many, many lives. He won't lose himself till he's ready to go, so to speak. And the doctrine of *karma* offers him a fairer life next time around if he treats people fairly this time through.

TRANSCENDENT BEING *AND* NATURAL WORLD

But Institutional Religions emphasizing a Transcendent Being can still accommodate something close to a Naturalist world view by positing a creation with the ability to organize itself—first into chemical elements, and eventually into replicating units that evolve by natural selection. A panpsychic account of consciousness could be included by providing the original creation with this fundamental property.

What about interventions of a Transcendent Being within the workings of nature so established? Well, unless a person agrees with Spinoza that the notion of a Transcendent Being outside of *"Deus sive Natura"* is logically untenable, then I fail to see why such interventions could not occur—just that such interventions would have to be conceptualized as *an aspect of one infinite act from all eternity,* rather than the step-by-step modifications provided from time to time by a humanoid god.

And once such interventions are accepted as logically possible, then so is life after death. For as we saw in Chapter 23, the *Star Trek* maneuver of reading out the type and position of each and every atom in a human body for perfect realignment at a later time is, in principle, possible. Voilá: *"resurrection of the body, and life everlasting, amen."* And *pari passu,* the deliverance of perfect justice.

WHY DO NATURALISTS NOT ACCEPT THIS MODEL?

Given then that a Supernaturalist model of the above sort does not seem to be incoherent, and given its advantage at a practical level of providing a happifying framework within which to conduct our lives—most especially when it comes to dealing with those core issues of *our insatiable longing for life and our unquenchable thirst for fairness*—why should Naturalists not join the majority of crew members on spaceship earth in joyfully grasping an explanatory story of this kind? Or to be more specific, why have I, over the course of my life, drifted away from such a Supernaturalist account?

I'd given some thought to addressing the issue here, but decided ultimately to leave that project for another day. The goal of this present book, after all, has been to describe some of the interesting mechanisms that underwrite our human view of the world, with all the beauty and limitations of that view. And since a satisfactory explanation for our human consciousness remains our most challenging intellectual frontier, the third and longest section of the book has been devoted to that particular issue. The goal of this final chapter then has been simply to address those readers who approach life from within a Supernaturalist frame of reference, pointing out that the Naturalist explanations provided herein can be assimilated to certain types of Supernaturalism, allowing one to avoid *"the necessity of postulating two incompatible truths."*

REFERENCES AND CHAPTER NOTES

PART I

CHAPTER 4:

1. Einstein, A. and Infield, L. (1966/1938) *The Evolution Of Physics*. Simon & Schuster: New York. (p 242)

CHAPTER 7:

1. Wilson, F.R. (1999) *The Hand*. Vintage Books: New York. (p 63)

2. In actual practice, we usually start with a fairly general label, then move to either more concrete or more abstract words from that middle position. For instance, we might start with the label *'tree'*; and as we learn to distinguish various types of trees, we then move to differentiating terms like *'oak tree'* and *'maple tree'*. We also might focus on what the trees have in common with other things, and then start to use more encompassing (more abstract) labels like *'plant'*.

CHAPTER 8:

1. Cohen, L.B., Rundell L.J., Spellman, B.A., and Cashon, C.H. *Infants' Perception of Causal Chains*. Available on web (cohen@psy.utex.edu)

2. Lerner, R.G., and Trigg, G.L., editors (1991) *Encyclopedia Of Physics*, VCH Publishers: New York.

CHAPTER 9:

1. This finding within our own grand and glorious species echoes at a more complex level the finding noted earlier in deerflies (Chapter 1) that *"when these insects are confronted with situations that call for cognitive skills beyond those naturally selected for coping with their native woodland surroundings, their systems fall into <u>error that is both systematic and predictable</u>."* We will address this general point again in Chapter 10, referring to it there as the *"Principle of Cognitive Limitation."*

Chapter 10:

1. As philosophers use the term 'knowledge', not only must one's judgment conform with what's actually the case, one must have good reason to so judge. For instance, if at a time when most medieval villagers believed that the earth was flat, one of their number insisted that the earth was round because, say, the thoughts in his own mind went round and round, his accidentally correct opinion, based on totally tangential reasoning, would not count as 'knowledge'.

CHAPTER 11:

1. DSM-IV (1994) (Diagnostic and Statistical Manual of Mental Disorders, Fourth Edition) American Psychiatric Association: Washington, DC.
2. Wittgenstein, L. (1953) *Philosophical Investigations.* Macmillan: New York.

PART II

CHAPTER 13:

1. Actually, Fabre's results were not always completely reproducible. That's because so many complex environmental factors can interfere with a given program's execution, and also (unusually) because DNA transcription can gets 'marred' by copying errors. However, such copying errors provide a species-upside over time because such variations will tend to be naturally selected if they give particular species-members an adaptive leg up.

CHAPTER 14:

1. Newberg, A., D'Aquili, E., and Rause, V. (2001) *Why God Won't Go Away.* Ballantine: New York.

CHAPTER 16:

1. Sullivan, P.R. (1995) The Natural Ought. *Behavior and Philosophy* 23: 1-12.
2. I simplify behaviorist terminology in this discussion by running together 'aversive reinforcement' (*e.g.* application of an electric shock to a rat's paws) and 'negative reinforcement' as the term is used technically (the withholding of 'posi-

tive reinforcement'). The distinction isn't necessary for our present purposes, especially since at the level of human behavior, *aversive* and *negative-in-the-technical-sense* often grade into one another. For instance, if a man who's coupled with a short-tempered woman fouls up, he can expect to be yelled at (aversive reinforcement). A man with the good fortune of living with a laid-back woman probably won't get yelled at, but he'll feel the temporary withdrawal of her smiling disposition as maybe even more painful; to wit, an "aversive reinforcement."

CHAPTER 17:

1. Bentham, J. (1988/1781) *The Principles Of Morals And Legislation.* Prometheus Books: Buffalo, NY.

2. Genesis 1:28 (<u>Douay</u> Version). Mention of the specific English translation emphasizes the problem that not only did humans have to decide which writings to accept as canonical, humans also had to do the translations (or as in the above version, the secondary translation from a particular Latin translation). Hence, literal interpretations of the Bible—of which our usually monoglot Fundamentalists are so fond—are confronted by an escalating series of difficulties.

CHAPTER 18:

1. Since almost any set of empirical results can be explained by more than one theory, there is no such thing—strictly speaking—as a "confirming experiment," only one that is happily congruent with the explanatory story being provided. In this sense, the only definitive result an experiment can provide is to *disconfirm* the explanatory story. And even here, in a large number of instances, the explanatory theory can be altered to accommodate the new findings. This may seem like cheating on a theorist's part, though we must bear in mind that scientific theories, in principle, are to be taken as *provisional;* that is, subject to the alternatives of either revision or rejection.

CHAPTER 19:

1. From Locke's "Letter on Toleration," quoted from Scharfstein, B. (1980) *The Philosophers.* Oxford University Press: New York. (p 163)

2. For additional details, Sullivan, P.R. (1995/1996) Murphy's Law and the Natural Ought. *Behavior and Philosophy* 223: 39-49.

CHAPTER 20:

1. At a less complex level, it seems likely that our human experience of orange, red, and yellow as "warm colors" comes from our response to fire and sun, while our human experience of blue as a "cool color" comes as a result of our contact with blue water.

PART III

CHAPTER 21:

1. Our brain-based information-processing systems also report sensations that become projected onto locations <u>within</u> the body, but there's quite a difference in localizing skill. To illustrate: If you experience a painful stimulation on the surface of your skin, you can localize the place where the stimulus is occurring with great accuracy. But if a noxious stimulus strikes in an internal organ, you're likely to experience the pangs in another location. Physicians call the phenomenon "referred pain," since the discomfort coming from, say, an inflamed appendix is "referred to" the umbilical area (only later, when the inflammation spreads to the abdominal lining, is the pain of acute appendicitis experienced in its classical "right lower quadrant" position near the actual location of the appendix).

Since it seems obviously advantageous for our central processing apparatus to project sensations to the actual location where stimulation occurs, the question arises: why did Natural Selection work so assiduously to correlate sensation with stimulus-location on the body's surface but not on the inside? One suggestion occurred to me as I was viewing a grim Civil War photograph of battlefield dead. Many corpses lay scattered about, and in the majority of cases, the dead soldiers' shirt tails were lying loosely outside their uniforms. What had given rise to this macabre detail, I wondered. Was it some passing fashion of soldiers going into combat that they left their shirts untucked? Was it the result of a callous frisking of their dead bodies for valuables?

The PBS commentator soon provided a factual explanation: Soldiers of the time knew that they were as good as dead once a missile penetrated their chest or abdomen, so when they were felled, they pulled out their shirt to learn the verdict. Lesson: in all eras before the availability of modern surgery and antibiotics, accurately locating internal lesions would have made virtually no survival difference; hence, there would've been little adaptive pressure to select for this additional ability.

CHAPTER 22:

1. Descartes, R. *Discourse On Method,* Part IV, paragraph 1.
2. ibid. paragraph 2.
3. Spinoza, B. The Ethics, Part V, Preface.

CHAPTER 23:

1. I confess to using the *"angels dancing on the head of a pin"* phrase with a hint of its usual derogatory implications—decrying the tendency of philosophers (of schools other than one's own, of course) to fritter away their time on precious academic issues irrelevant to the enterprise of human life. Current Scholastics would reply, however, that this notorious example, if it indeed ever actually occupied the time of Scholastics, amounted to what would be referred to in current parlance as a "thought experiment." Specifically, how is one to relate an object that has no extension (*e.g.* an angel) in any meaningful way to our everyday world of physical dimensions?

2. In the scientific manner of thinking, we usually find multiple hypotheses, and this particular instance can stand as a good example of that fact. The author of my college textbook of genetics [Snyder, L. H. (1951) *The Principles of Heredity.* Boston: D.C. Heath] favored a different explanation, and he felt strongly enough that he represented it as a fact: *"We shall discover that hereditary factors are <u>complex protein molecules,</u> which we call genes. Most of the genes are located at definite points in microscopic structures, called chromosomes…"* His opinion was shared by many scientists before Watson and Crick unraveled the double helix of DNA, because they thought only proteins possessed sufficient complexity to contain the vast amount of information that needed to be passed on genetically.

3. Luke, 23:43.

CHAPTER 25:

1. Dennett, D.C. (1991) *Consciousness Explained.* Little, Brown and Company: Boston. (p 433)

2. Singer, T., Seymour, B., O'Doherty, J., Kaube, H., Dolan, R.J., and Frith, C.D. (2004) Empathy for Pain Involves the Affective but not Sensory Components of Pain. *Science* 303: 1157-66.

3. McGinn, C. (1993) *The Problem of Consciousness.* Blackwell: Cambridge.

CHAPTER 26:

1. This instance provides us with a good opportunity to illustrate how 'hard-wired' brain programs are harnessed in various ways on the basis of experience. The basics of this motor program can be seen in an infant a few months old: If, while he is lying on his back, a doctor rotates his head, say, to the left, the baby's left arm and leg will extend, and the opposite limbs will flex. After several additional months of life, this reflex will cease to occur in response to head rotation—not that the motor program involved has vanished; it's become inhibited by the development of higher centers, so it doesn't show itself automatically in response to the head-rotating stimulus. But the program still shows itself in a variety of bodily movements through life, as can be noted readily in athletic activities (*e.g.* a sudden cut by a running back in football or a driving guard in basketball).

2. Watson, J.B. (1924/1970) *Behaviorism.* New York: W.W. Norton. (p 2)

3. ibid. p 3.

4. Searle, J.R. (1992) *The Rediscovery Of The Mind.* MIT Press: Cambridge (p 5)

CHAPTER 27:

1. For instance, once Aristotle got to clicking off all the properties shown by a particular *type* of substance, it was easy to then group (A) the properties that a given substance needed to have in order for it to *be* the kind of substance it was (*e.g.* a 'canoe' must be tapered), and (B) the properties that might well be present, though they wouldn't have to be (*e.g.* a canoe's yellow color). The first properties were essential, whereas the second properties were merely incidental ("accidental," as the Scholastics would say). Focusing then on the first group of properties that were *always and necessarily* present in a given *type* of substance, Aristotle eventually came to use the notion of 'substance' in the sense of 'essence'.

2. Locke, J. *An Essay Concerning Human Understanding:* Chapter 23, paragraph 15.

3. Spinoza, B. The Ethics, Part II, Prop. XIII, Note.

4. ibid.

CHAPTER 28:

1. Weinberg, S. (1992) *Dreams Of A Final Theory,* Vintage Books: New York. (p 45).

2. Lucretius, *On The Nature Of The Universe,* translated by R.E. Latham (1951) New York: Penguin Books. (p 28).

CHAPTER 29:

1. Holmes, O.W. (1881/1963) *The Common Law.* Cambridge: Harvard University Press. (p 10-11).

2. ibid. (p 12).

3. Chalmers, D.J. (1996) *The Conscious Mind.* Oxford University Press: New York. (p 168).

4. Searle, J.R. (1997) *The Mystery of Consciousness.* New York Review of Books: New York. (p 161).

5. ibid. (p 166).

6. ibid. (p 170).

CHAPTER 30:

1. Genesis 18, 23-26

2. Genesis 18, 28-30

3. Note, for instance, that Abraham is suggesting to God that He is being unfair to the just, and that *"This is not beseeming thee."* But here's the irony: In the initial assumption of Supernaturalist morality, right and wrong are to be based solely on the Will of God; it is only right that we obey *His* Laws. But in accordance with this principle, if God chooses to modify His Laws, that's His business, so to speak. (Folks certainly don't complain when, say, He abrogates His laws in order to perform health-giving miracles at Lourdes.) But in the present instance, it's as if Abraham is saying to God: *"You've placed the laws of justice in our hearts; now You can't go changing them just to suit Yourself; that ain't fair!"* In other words, Abraham seems to be judging God, not by starting with a Supernaturalist assumption—*"Thy will be done"*—but by our own human standard of what constitutes fairness. And Abraham's ancient standard, we might add, continues to seem right to us; for in a far different time and place, we also don't think it fair to punish an individual for a crime he hasn't committed.

4. Spinoza, B. (Letter XXXVI).

5. Voltaire (1989/1736) "A Treatise on Metaphysics," in *Voltaire Selections.* New York: MacMillan. (p 93).

6. More familiar to current readers might be C.S. Lewis' well known book, *Mere Christianity* [(1952) Macmillan: New York], which uses a variation of the same "argument from design"—starting with our inborn sense-of-fairness, and asserting that the presence of this innate principle requires a Designer (Law Giver) to have instilled it within us.

7. Dawkins, R. (1996) *The Blind Watchmaker.* W.W. Norton: New York.

8. Weinberg, S. (1992) *Dreams Of A Final Theory,* New York: Vintage Books. (p 245).

9. ibid. (p 250).

CHAPTER 31:

1. Nagel, T. (1979) What Is It Like To Be A Bat?, in *Mortal Questions.* Cambridge University Press: New York. (p 168).

2. ibid. (p 168).

3. Turing, A.M. (1950) Computing Machinery and Intelligence. *Mind* 49: 433-

4. Quotation referenced in Radner, D. and M. (1996) *Animal Consciousness.* Prometheus Books: Amherst, NY. (p 60).

CHAPTER 32:

1. Dennett, D.C. (1991) *Consciousness Explained.* Boston: Little, Brown and Company. (p 418).

CHAPTER 33:

1. Greenfield, S.A. (1995) *Journey to the Center of the Mind.* W.H. Freeman and Company: New York. (p 91).

2. For additional details, see Sullivan, P.R. (1995) Contentless Consciousness And Information-Processing Theories Of Mind. *Philosophy, Psychiatry, & Psychology.* 2: 51-59, and Binns, P. (1995) Commentary on Contentless Consciousness, *Philosophy, Psychiatry, & Psychology.* 2: 51-59.

3. Searle, J.R. (1992) *The Rediscovery Of The Mind.* Cambridge: MIT Press. (p 70).

4. ibid. (p 84).

5. Lycan, W.G. (1996) *Consciousness and Experience,* Cambridge: MIT Press. (p 164). Lycan also provides a helpful description of 8+ ways in which the term 'consciousness' has been defined.

6. Searle, J.R. (1997). (p 112).

7. James, W. (1890/1950) *The Principles Of Psychology* (Vol 1). Dover Publications: New York. (p 133).

8. Stace, W.T. (1960/1987) *Mysticism And Philosophy.* Los Angeles: Jeremy Tarcher. (p 110).

9. Interview available on internet (Google) under 'pure consciousness', John Horgan.

CHAPTER 34:

1. Sometimes, an experiment that is possible in principle is not doable in practice because of *technical limitations*. For instance, existing particle accelerators limit the actual force attainable during particle collisions. In the present instance, not technical limitations but rather *ethical problems* might well prevent the specific experiment, since prospective participants would obviously not be in position to provide 'informed consent'. However, there are other experimental methods that have been used to produce syncope (e.g. the tilt table), and these might be effective in producing the type of unconsciousness needed (c.f. the report by Herzon noted earlier). If so, volunteers could be ethically solicited.

2. In accordance with Murphy's Law, desynchronization can also occur in the opposite direction. Hence, instances occur in which muscle systems *don't* get decommissioned during a dream—occasionally resulting in significant injury.

3. As I'm using the phrase here, *"Outside our ordinary human world"* refers to an event that is not experienceable when our human machinery is operating within its normal channels, but with no implication that the event is outside of nature itself. The event is <u>not</u> supernatural. Notice then that I'm analyzing the event from within both the perspective of my normal human cognitive apparatus AND from within a set of Naturalist assumptions. Many religious mystics, when analyzing pure consciousness events, do so from within a set of Supernaturalist assumptions; hence, while they also refer to such events as *"outside* our ordinary human world,"* these commentators are referring to what they believe to be supernatural phenomena.

4. Spinoza, B. *The Ethics,* Part V, Prop. XXIV.

5. From *Auguries of Innocence* c. 1800.

CHAPTER 35:

1. Let's use an illustration to emphasize this point. Bernard Baars has provided an account of a crucial feature that's necessary in the elaboration of human consciousness: <u>Information that gets involved in conscious experience must be of a sort that's widely broadcast throughout the brain</u>, hence available for diverse uses by many neural systems. When information is present, but the information is not available widely enough and within the proper systems, afflicted patients may still be able to use the information in certain ways. For instance, patients with so-called 'blindsight' may correctly guess, *if asked,* whether a stimulus has been turned on in the blind part of their visual fields. However, they will have no visual awareness of the stimulus, and will have a sense that they're doing no more than guessing.

While Baars' account is helpful because it highlights a crucial element in those brain operations associated with consciousness, the account doesn't explain why human consciousness must inexorably occur under such circumstances, rather than simply broad-based information-processing without consciousenss. Broad-based sharing of information is clearly necessary if analyzing and weighting systems within the human brain are to produce the wide range of decisions necessary for adaptive behavior. But if the underlying neural networks were doing their same job without conscious accompaniment, would they not result in the same behavioral output? [Baars details his thesis in (1988) *A Cognitive Theory of Consciousness.* Cambridge University Press: New York.]

2. A sense of the tenacity with which we humans cling to notions like the continuation of life after death can be gotten from noting the vehemence with which Spinoza was drummed out of his community [as quoted from Nadler, S. (1999) *Spinoza, A Life.* Cambridge University Press: New York]:

> ...Cursed be he by day and cursed be he by night; cursed be he when he lies down and cursed be he when he rises up. Cursed be he when he goes out and cursed be he when he comes in. The Lord will not spare him, but then the anger of the Lord and his jealousy shall smoke against that man, and all the curses that are written in this book shall lie upon him, and the Lord shall blot out his name from under heaven...

Looks like the edict didn't leave that wicked desperado, Spinoza, much wiggle room. Fortunately, as my mother used to say: *"Sticks and stones may break my bones, but names will never hurt me."*

0-595-34602-2

www.ingramcontent.com/pod-product-compliance
Lightning Source LLC
Chambersburg PA
CBHW020733180526
45163CB00001B/225